The Moving Spotlight

The Moving Spotlight

An Essay on Time and Ontology

Ross P. Cameron

OXFORD
UNIVERSITY PRESS

Great Clarendon Street, Oxford, OX2 6DP,
United Kingdom

Oxford University Press is a department of the University of Oxford.
It furthers the University's objective of excellence in research, scholarship,
and education by publishing worldwide. Oxford is a registered trade mark of
Oxford University Press in the UK and in certain other countries

First Edition published in 2015

Impression: 1

Published in the United States of America by Oxford University Press
198 Madison Avenue, New York, NY 10016, United States of America

British Library Cataloguing in Publication Data
Data available

Library of Congress Control Number: 2015933915

ISBN 978–0–19–871329–6

Printed and bound by
CPI Group (UK) Ltd, Croydon, CR0 4YY

For my girls, Elizabeth and Willow

Acknowledgments

Thanks to David Armstrong, Karen Bennett, Phillip Bricker, Rachael Briggs, Andrew Cortens, Shamik Dasgupta, Louis deRosset, Cian Dorr, Antony Eagle, Kenny Easwaran, Maya Eddon, Andy Egan, Matti Eklund, Kit Fine, Brie Gertler, Caspar Hare, Katherine Hawley, John Hawthorne, Hud Hudson, Dave Ingram, Jonathan Jacobs, Carrie Jenkins, Jeff King, Heather Logue, David Manley, Ned Markosian, Kris McDaniel, Trenton Merricks, Sarah Moss, Daniel Nolan, Jill North, Josh Parsons, Laurie Paul, Graham Priest, Agustin Rayo, Mike Rea, Jeff Russell, Jonathan Schaffer, Ted Sider, Brad Skow, Jason Stanley, Helen Steward, Meghan Sullivan, Jonathan Tallant, Christina Van Dyke, Meg Wallace, Timothy Williamson, Jessica Wilson, Richard Woodward, Crispin Wright, Dean Zimmerman, and the graduate students of the universities of Virginia and Leeds. Thanks also to two anonymous referees for Oxford University Press, and to Peter Momtchiloff for his excellent editorial work, advice, and encouragement.

Special thanks to Jason Turner. I was fortunate enough to be colleagues with Jason for several years at Leeds, and my many enjoyable philosophical conversations with him have significantly shaped my philosophical outlook.

Very special thanks to Robbie Williams. Robbie and I were colleagues at Leeds for almost a decade, and graduate students at St Andrews for three years before that. There's not a philosophical issue on which my thoughts haven't been shaped by our many wonderful conversations, to the extent that it's hard for me to tell just where my thoughts end and his begin. I can't really conceive what life would be like now if we hadn't "grown up" together, but I know it would be much worse. Needless to say, any mistakes that remain in this book are his fault.

Mega uber-special ultra-thanks to Elizabeth Barnes, my wife, colleague, favorite philosopher, and best friend. She has helped me with every part of this book: not a single idea in it hasn't been improved by my conversations with her, and she has been a source of support and encouragement whenever I got disheartened. Without her, nothing would be worth anything.

Most of the material in this book is previously unpublished, but some parts are parts of previously published papers, for which I give the following thanks.

Sections 4.2 and 4.3 are lightly adapted from part of my paper "Truthmaking for Presentists," *Oxford Studies in Metaphysics*, Vol. 6, edited by Karen Bennett and Dean Zimmerman, Oxford University Press (2011, 55–100). Thanks to Oxford University Press for permission to re-use this material.

Section 4.6 was (with minor differences) originally published as "Changing Truthmakers: Reply to Tallant and Ingram," *Oxford Studies in Metaphysics*, Vol. 8, edited by Karen Bennett and Dean Zimmerman, Oxford University Press (2013, 362–73). Thanks to Oxford University Press for permission to re-use this material.

Section 5.2 is taken from part of Elizabeth Barnes and Ross Cameron, "Back to the Open Future," *Philosophical Perspectives* Vol. 25, *Metaphysics* (2011, 1–26). Thanks to John Wiley & Sons Ltd. for permission to re-use this material. Thanks also to Elizabeth Barnes for her consent to have this part of our co-authored paper re-used here.

Contents

Introduction

Before I begin I want to say something about how this book is written. I think there are two ways to write a philosophy book: start out by stating the main theses to be argued for, and then take your reader through the arguments for those theses; or take your readers through the arguments, *revealing* your theses as you go. Stylistically, I like the latter. The book becomes like a detective story, the arguments the clues, and the theses the big reveal. And just as with a detective story, you hope that the conclusion seems both surprising and yet inevitable. However, pulling in the other direction is the fact that philosophical arguments are often easier to follow if you know what the author is arguing for. What I would *really* like is for you to read the book, discover the conclusion, and then go back and re-read it to re-evaluate the arguments with that conclusion in mind. This may be wishful thinking. And so I have opted for a compromise. I am not going to state the main theses to be argued for at the outset, for those like me who enjoy a dose of *whodunnit* with their philosophy. But there are two sections that if you *want* a spoiler you can go and read now, and then come back and work through the arguments: they are the list of the main theses of my proposed metaphysic in section 4.8 (section 4.7 might also be useful), and the conclusion at the very end of the book. If you want a sense of what I am building towards as you read me doing it, go and read them now, then come back. Done? Then let's begin.

The project of this book is to defend what I think is the best A-Theoretic metaphysic: a version of the moving spotlight view. You will not find much in the way of argument *for* the A-Theory in these pages; rather, the intention is to assume that some such version of the A-Theory is the correct metaphysic of time, and to ask on that assumption what the best version of the A-Theory is. (The only exception is that in section 4.5 I will argue that in order to be an endurantist you must be an A-Theorist. So insofar as you are prepared to accept "Endurantism is true" as a premise, this is an argument for the A-Theory.) In that case, I should start by making the assumption clear: what do I mean by the claim that some version of the A-Theory is the correct metaphysic of time?

I mean the combination of two claims. The first claim is:

> *Privileged Present*: There is a unique objectively privileged time: the time which is present. No description of reality can be correct and complete without specifying which time is present.

The second claim is:

> *Temporary Presentness*: What time is objectively privileged changes: the time that is objectively present either was or will not be present (or both), and some time that is not objectively present either was or will be (or both).

I will be using the term "A-Theory" such that it is necessary and sufficient for a theory to be an A-Theory that it entail both *Privileged Present* and *Temporary Presentness*. This is purely stipulative: the term is not used uniformly in the literature to determinately single out any unique set of doctrines, so we are left with some amount of choice in how to use the term, and this is my choice.

One could be more restrictive, and use "A-Theory" just to mean those theories that entail *Privileged Present*. I choose not to do so simply because theories on which *Privileged Present* is true but *Temporary Presentness* false are not very interesting. Here is a view: the stuck spotlight theory—all times (past, present, and future) are real, and there is an objectively privileged time, and it is always November 30, 1982. That is a coherent view; for all I know, it is even metaphysically possible. But it is hard to see how we could have any reason to believe in such a view. It would be perverse to call the objectively privileged time "the present," because that would make a mockery of our tensed talk. Assuming that talk about what is happening now is talk about what is happening in the present, this view entails that "Michael Jackson's album *Thriller* was released today" always has been and always will be true. That would render talking about what is happening now useless; if we found ourselves in the stuck spotlight world it would be better to simply use tensed language as the B-Theorist does, and take talk of what is happening now to just be talk of what is happening simultaneously with the utterance. And that is fine: we can talk like B-Theorists even if there is an objectively privileged time—we are the masters of our words, after all. But in that case, we should not call the privileged time "the present"—we should agree with the B-Theorist that "the present" just marks out our temporal location like "here" marks out our spatial location: A's utterance of "t is present" is true iff A's utterance occurs at t. So there is no privileged *present* on the stuck spotlight view: it is just that something is objectively privileged, and it happens to be a time. But the privileged status of this time seems to be purely epiphenomenal, doing no work at all. Why on Earth would anyone ever believe such a theory, then? We wouldn't: it is of no interest.

And so we need to add *Temporary Presentness* to get an A-Theory. Note that this claim is far weaker than what we expect to be true concerning changes in presentness. It allows that there are many times that are not, never were, and never will be present. All it demands is that one time is present and then another different time is present. The A-Theorist is likely to hold something much stronger: that *every* time was, is, or will be present, and that the present moves along from one time to another in a principled manner: that is, that time t1 is present, then t2, then t3, etc., etc., rather than time t1 being present, then t2, then t1 again, after which the present jumps to t3. But while these claims are likely to be held by the A-Theorist, I hesitate to stipulate that they are constitutive of being an A-Theorist. Someone who thinks that there is a privileged present that behaves in a weird manner—so that some time never was and never will be present, say—might be being perverse, but I think they have a view on the metaphysics of time that is close enough to my own to make it worth thinking of the two views as of a kind. So while *Temporary Presentness* is weak, I will consider it all that is needed in addition to *Privileged Present* in order to get an A-Theory.

So much for what the A-Theory *is*; let me make some clarificatory remarks on what the A-Theory, as I understand it at least, is *not*.

The A-Theory is not a theory about tensed language or thought: it is a theory about *reality*, not our *representations* of reality, be they linguistic or mental representations. And so the A-Theory is not the theory that tensed language or thought—by which I mean language or thought that makes essential use of A-Theoretic vocabulary like "is present," "is past," "is future," etc.—is ineliminable. What *is* the claim that tensed language or thought is ineliminable? On one reading, *all* language and thought is eliminable, for we can just stop talking and thinking. But that no language or thought is ineliminable in this sense is uninteresting. More interesting would be the claim that while you can, of course, simply refrain from using tensed language or thought, you would thereby inevitably be missing out on some feature of reality: you would inevitably fail to be able to represent some way reality is. Perhaps, then, the claim that tensed language or thought is ineliminable is meant to be the claim that any complete and correct linguistic or mental representation of reality must make use of tensed language or thought. Understood thus, the A-Theory will entail that tensed language and thought is ineliminable *if* it entails that the A-properties are irreducible. For the A-Theory entails that there is a unique present time; if it is true, then, any complete and correct representation of reality has to represent there being a unique present time; if the A-property *is present* is irreducible, then the only way for a linguistic or mental representation to represent that there is a unique present time is by employing A-Theoretical language or thoughts; thus, if the A-properties are irreducible, then tensed language or thought is ineliminable in this sense: it must be invoked to give a complete and correct linguistic or mental representation of reality. However, that is a big "if": the A-Theorist may well hold that the A-properties are irreducible, but as we will shortly see, they need not do so. And if the A-properties can be reduced, then one can have a complete and correct linguistic or mental representation of reality that eschews tensed language and thought, and only invokes that language or those thoughts that are required to represent how things stand at the reductive base. So this claim that tensed language or thought is ineliminable, while it may well be held by many an A-Theorist, is not part of what it is for a theory to be an A-Theory. Nor is the A-Theory to do with the ineliminability of tensed language or thought in the sense that such language or thoughts play a crucial role that cannot be played by untensed language or thoughts, for the B-Theorist can happily agree with that claim, just as those impressed with John Perry's case of the essential indexical can think that first-person language is ineliminable, in that sense, without being committed to an underlying domain of first-personal facts in reality.[1]

The A-Theory is not a theory about the irreducibility of the A-properties: that is, the properties *being present, being past*, and *being future*. The A-Theorist will not of course accept the B-Theoretic reduction of those properties as *being simultaneous with this thought, being before this thought*, and *being after this thought*; but she need not commit to there being no such reduction. It is perfectly consistent with the A-Theory that what it is to be present is to be the object of God's attention, with what it is to be past being what is before the object of God's attention and what is future being what is after the object of God's attention. I do not recommend such a reduction, but I do not think the believer in it thereby disqualifies herself from being an A-Theorist.

The A-Theory is not the claim that there is a metaphysical difference between time and space. With a certain plausible auxiliary assumption—namely, that there is no privileged spatial location—the A-Theory will *entail* that there is a metaphysical difference between time and space. But that is an additional assumption: one who held the perverse view that there *is* a privileged spatial location—the top of Arthur's Seat in Edinburgh, say—as well as holding that there is a privileged temporal location—the present—would still be an A-Theorist about time, but would deny that there is this metaphysical difference between time and space. And certainly, this thesis is not sufficient to be an A-Theorist. Maudlin thinks that there is a metaphysical difference between time and

[1] Perry (1979).

space: time, but not space, has an intrinsic direction.[2] But Maudlin is not an A-Theorist, as I understand it: for while he believes in an arrow of time, he does not believe in a privileged moment in time.

The A-Theory is not a thesis about truth, nor is it a thesis about how truths correspond to reality. In particular, it is not the thesis that propositions do not have truth-values *simpliciter* but rather only truth-values *at a time*.[3] As I see it, the A-Theorist and B-Theorist need not disagree about the nature of truth. Both of them can think that propositions are true or false *simpliciter*; and both of them can think that a proposition is true iff it corresponds, simpliciter, to how reality is. The difference arises because the A-Theorist thinks that some of the propositions that are true/false simpliciter *were* and *will be* false/true simpliciter. And this happens because how reality is *changes*, so whether or not a proposition corresponds (simpliciter) to reality changes,[4] leading to a change in truth-value. (All this will be gone into in more detail in chapter four.)

What is essential to the A-Theory is that things *change*. One way in which things change is that things change their properties: I am 6ft tall, but I used to be 4ft tall. But this does not force us into saying that nothing is 6ft tall simpliciter, merely 6ft *at* some times and 4ft *at* other times. We need not reject the dyadic relation of instantiation that holds between an object and a property in favor of a triadic one that holds between an object, a property, and a time. All we need say is that I am 6ft tall, simpliciter, but that I was 4ft tall simpliciter. (This will become important in chapter four.) The dyadic relation of instantiation holds between me and the property of being 6ft tall. End of story. However, things *were* different: it used to be the case that the dyadic relation of instantiation held between me and the property of being 4ft tall instead.

Being true is just another property, one that can be had or lacked by propositions. And when it is had/lacked, it is had/lacked simpliciter. But that is completely compatible with it being the case that this property is had temporarily, and that things were different with respect to a proposition's being true.[5] The fact that

[2] Maudlin (2007, ch.4).

[3] *Contra* Ned Markosian, who says "[W]e'll assume the 'tensed' conception of semantics, according to which the bearers of truth and falsity are to be assigned truth values at times, and also according to which the past and future tenses are ineliminable features of language. (I take both components of this tensed conception of semantics to be required for Presentists, and indeed for anyone who endorses The A Theory of time.)" Markosian (2014b, p132).

[4] Thus I deny Markosian's (1995) claim that the A-Theorist need renounce the traditional correspondence theory of truth in favor of one that talks not of propositions corresponding to reality, simpliciter, but rather corresponding to reality *at a time*.

[5] Cf. Cappelen and Hawthorne (2010).

things change—that what is true used to and will be false—does not force us to reject truth simpliciter in favor of truth at a time, it simply forces us to say that one of the ways in which things change is the having of the monadic property of truth by a proposition. Similarly, *correspondence* is just another relation, one that can hold or not between a proposition and the world. And either it holds between the proposition p, say, and the world, or it does not, simpliciter. But of course, that may change: p may correspond to the world, simpliciter, but then later this relation may fail to hold simpliciter between those two relata. This is no more puzzling than the fact that the dyadic relation *is taller than* holds between me and my mother, simpliciter, but didn't used to. In each case, the relation holds temporarily due to one or more of the relata changing. The *is taller than* relation holds between me and my mother but used to hold the other way around—and that is because I changed: I grew taller over time. The correspondence relation used to hold between p and reality but does so no longer—and that is because reality changed: it used to make p true, but now does not. We do not need to reject the dyadic correspondence relation in favor of the triadic correspondence at a time relation; we simply need to allow that whether things are related by a relation is one of the features of reality that are subject to change. Likewise, we do not need to reject the monadic notion of a proposition's being true in favor of the dyadic notion of a proposition's being true at a time; we simply need to allow that whether a proposition has that property is one of the features of reality that is subject to change.

So much for what the A-Theory is and is not. Here are three familiar ontological options for the A-Theorist:

Presentism: The only concrete substances that exist simpliciter are present entities.[6]

The Growing Block: Every concrete substance that did exist or that presently exists, exists simpliciter. No concrete substance that merely will exist exists simpliciter.[7]

The Moving Spotlight: Every concrete substance that did exist, or that exists now, or that will exist, exists simpliciter.[8]

Each view thinks that temporal ontology is more expansive than the previous view: the presentist limits their temporal ontology to present beings, whilst the growing blocker includes past beings in addition, and the moving spotlighter includes both past and future beings.

[6] For a defense, read Bigelow (1996), Markosian (2004), and Merricks (2009, ch.6).
[7] For a defense, read Broad (1923) and Tooley (1997).
[8] For a defense, keep reading this book.

I have cast these views in terms of what concrete substances exist. The point of the concreteness criterion is to limit our attention to things that exist in time. As I see it, it is compatible with presentism that there exist some things that are not present entities, because they simply do not exist in time at all—such as numbers, or (on some views) God. I shall stipulatively call things that exist outside of time abstract, and limit the scope of the various A-Theoretic claims to concreta. What is essential to presentism is simply that it not posit any non-present concrete substances like dinosaurs or lunar colonies. The point of restricting the scope to concrete *substances* is to focus our attention on particulars rather than states of affairs involving them. The moving spotlight is often stated as the view that there is no change with respect to what is in the domain of the unrestricted quantifier: that anything that did or will exist always exists simpliciter. But as we will see in chapter four, the version of the moving spotlight theory I will develop actually allows that there are some changes in what there unrestrictedly is, for states of affairs that did exist in the past exist no longer, and states of affairs that will exist do not exist yet, etc. What states of affairs unrestrictedly exist changes, because how things are changes. But the view is a version of the moving spotlight, as I will be using the term, because what concrete *substances* there unrestrictedly are—what particulars are around to be the *constituents* of states of affairs—does not change. Caesar, myself, and the first lunar colony always exist, but how we are—and therefore what states of affairs involving us unrestrictedly exist—changes.

In stating these three theories, we presuppose that there is a sensible notion of existence *simpliciter*. Some philosophers[9] are inclined to deny this, thinking that the only notions of existence we have are tensed: there is what exists *now*, on the one hand, and, on the other, there is what exists now, or did exist, or will exist. And of course, it is entirely trivial that only present entities exist *now* and that past, present, and future entities exist now, or did exist, or will exist; so on this view, there is simply no interesting question concerning the extent of temporal ontology. I find this view very counter-intuitive: just as (or so it seems to me) I can simply ask whether there are numbers, or Gods, or countries, or electrons, so I can simply ask whether there are dinosaurs, or lunar colonies, without qualifying the notion of "existence" in question. There is what there is, simpliciter, and I can ask whether amongst those things there are non-present entities. And so, throughout this book I will assume it makes sense to speak of existence, simpliciter, and that there is a substantive question as to whether amongst the things that exist, simpliciter, there are non-present entities.

[9] See e.g. Meyer (2013).

I.1 The Plan for this Book

This book will defend a version of the moving spotlight theory, although it will be a version of the moving spotlight theory that shares with presentism the thesis that the way things are now is the way things are simpliciter.

The moving spotlight view has not been widely popular, and I think it is easy to see why. If you look at the familiar problems and issues that are discussed in the metaphysics of time literature, the natural response to some of them is that presentism comes out best, to others that the growing block view comes out best, but the moving spotlight view never seems to do better than its two main rivals. Let us look briefly at some of these familiar problems.

I.1.1 The Epistemic Problem

How can you know that you are present? For the B-Theorist, there is no puzzle, due to the indexical nature of "present" on their account. For the B-Theorist, if I utter "I am in the present," my usage of "the present" merely refers to the time of the utterance, just like when I utter "I am here" my usage of "here" refers to the place I am when I utter it. And so I am guaranteed to be present just as I am guaranteed to be here. But for the A-Theorist, "the present" does not simply refer to the time at which the utterance is located, it refers to the objectively privileged time. So how can I know that it is the time at which I make the utterance that is objectively privileged? How can I know that it is this time that is the objective present?

The orthodox view here[10] is that this is a serious problem for non-presentist A-Theoretic metaphysics like the growing block or the moving spotlight, but that it is no problem for presentism. Presentism says that there only *is* one time, the present. So, of course, it is this time that is present—that's the only time there is! However, if there are many times then the problem is vivid: how do I know that out of all the times there are that this one is the privileged one? For the growing blocker the problem is: how do I know I am at the *edge* of the block, rather than buried in the middle of the block (in the past)? For the moving spotlighter the problem is: how do I know the spotlight is on me as opposed to one of the many times before me (putting me in the future) or one of the many times after me (putting me in the past)?

So the natural conclusion here is: presentism is okay, but the growing block and the moving spotlight both (and to the same extent) have a problem.

[10] As in Bourne (2002, 2006), Braddon-Mitchell (2004), Heathwood (2005), and Merricks (2006).

I.1.2 McTaggart's Paradox

McTaggart (1908) argued that time was unreal because time required there to be an A-series of events and that results in an inconsistency. Nobody was convinced that time was unreal, but some people[11] were convinced that McTaggart was right to conclude that the A-series was inconsistent, and they concluded that all A-Theories must be false: presentness, pastness, and futurity must not be intrinsic properties of times and things in time as the A-Theory says but rather must simply concern our relation to things in time, as the B-Theory says.

A-Theorists, of course, must reject this conclusion. And some[12] think that McTaggart's argument does not damage any version of the A-Theory. But others[13] think that McTaggart's argument *does* pose a problem for non-presentist versions of the A-Theory but not for presentism. This latter conclusion is easy to reach. Jettisoning the charges of circularity and of infinite regress, McTaggart's argument in its simplest form can be seen as demonstrating that the following tetrad is inconsistent:

1. Event E is past
2. Event E was present
3. If something was F then it is F in the past
4. Nothing is both past and present

The argument for the inconsistency is as follows. From (2), we learn that there is an intrinsic property—*being present*—that E used to have. Item (3) then tells us that it is the case—albeit that it is the case in the past—that E is present. But the only location E has is in the past, so it seems that E must simply be present, simpliciter. Number (1) tells us that E is past, so E is both past and present; but that conflicts with (4).

It is easy to see why the presentist need not be worried about this version of McTaggart's argument: she will reject (3). For the presentist, it can be true that something was F without there being any reality at all to that thing's being F. Reality is exhausted by present reality, thinks the presentist: so things that merely were the case have no reality whatsoever. In particular, they are not the case in the past, for there is no past.

But there is a *prima facie* problem here for the growing block and moving spotlight theories. They do believe that the past is real. The things that happened

[11] Such as Mellor (1998).

[12] Zimmerman (2007) calls arguments based on McTaggart's one "specious" (p216) and "singularly unimpressive" (p216, fn.7).

[13] Such as Craig (1998).

are just as real as the things that are presently happening. On the face of it, then, they will accept (3): the way things were *is* the case, but in the past. And that is what gets McTaggart's argument going.

So as with the epistemic problem, it is natural to think that McTaggart's paradox poses a problem for non-presentist A-Theories, but that presentism is immune.

I.1.3 Truthmaking

It is a datum that there are true claims about what happened. (Whereas the claim that there are true claims about what will happen is, perhaps, negotiable: see the discussion of the open future below.) In virtue of what are these claims about what happened true? The growing blocker and the moving spotlighter have an easy answer, seemingly: in virtue of the existence and features of past ontology. There is a past, and it is a certain way, and it is the past being the way it is that grounds the truth of true claims about what happened. The presentist obviously cannot say this, since they do not believe in the past. And, familiarly, they have a hard time accounting for true historical claims since they can only appeal to things that exist now, and the properties they now have, and *prima facie* the nature of present reality leaves underdetermined how things were. Of course, the presentist *can* complicate present ontology such that there is not this underdetermination: they can, for example, postulate primitive tensed properties that things have now and that ground truths about how they were.[14] But many[15] have claimed that such a move is objectionable on other grounds. The threatened conclusion is that there is no way for the presentist to admit grounds of historical truths without believing in something that is independently objectionable.

So here the natural conclusion is that presentism faces a problem, but that the growing block and the moving spotlight theory do not.

I.1.4 Relations between Present and Non-present Things

On the face of it, things that exist now stand in lots of relations to things that did exist but that do not exist now. I admire Leibniz; the window is now breaking because of the past event of my throwing a brick at it. It seems plausible that for at least some of these relations, the relation cannot hold without the relata existing. Thus there is pressure to believe in the past entities

[14] Such as in Bigelow (1996).
[15] See *inter alia* Sider (2003a, p185), Merricks (2007, p135), Caplan and Sanson (2010).

to which present entities are related. No problem for the growing block or moving spotlight theories, but a problem for presentism, given that it says these past entities do not exist.

And as with the previous discussion of truthmakers, it is less obvious that present things stand in relations to things that will exist but that do not exist now. (We will come to whether we ought to hang on to this thought below.) So the natural conclusion here, again, is that presentism faces a problem, but that the growing block and the moving spotlight theory do not.

I.1.5 The Open Future

Intuitively, there is a metaphysical difference between the future and the past: the past is fixed but the future open. There is one fixed way that history was, but there are many ways that the future might unfold, or so at least many have thought. Assuming there is this metaphysical asymmetry between the past and the future, *why* is there such an asymmetry? The growing block theory offers us an answer: this metaphysical asymmetry is grounded in an ontological asymmetry. It is because the past is real that claims about what happened are fixed, and because the future is unreal that claims about what will happen are open. However, both presentism and the moving spotlight theory hold that the past and future are on a par ontologically: equally real according to the moving spotlight, equally unreal according to presentism. This threatens to rule out—or at least render mysterious—their not being on a par metaphysically.

Here, then, the natural conclusion is that presentism and the moving spotlight theory face a problem, but that the growing block view does not.

I.1.6 What Is It *for the Present to be Privileged?*

The A-Theory says that the present is metaphysically privileged. But what does that mean? What *is it* for the present to be privileged? The present time is the one that claims about what is happening *now* are sensitive to. But it does not seem that this is *what it is* for the present to be privileged: rather, it is *because* the present is privileged that claims about what is happening now are sensitive to what's happening at it. So there must be some other account of what makes that time special.

The presentist has such an account: the present is privileged because it is real. The present is the sum total of reality: it encompasses all that there is. Claims about what is happening now are sensitive to what is happening in the present because present reality just is reality, and claims about what is happening now are just claims about what is happening.

The growing blocker also has an account: the present is privileged because it is the edge of being. It is the unique time such that nothing is beyond it. Now, of course, there are two edges of being, if the past is finite; but even so, there is a clear reason to treat one edge as special—it is the edge that will be added to. So the present is privileged because it is the edge that just came into being: it is the unique time that represents the last point of reality as it is up until now. Claims about what is happening now are sensitive to what is happening at the present because claims about what is happening now are claims about the latest addition to reality.

The moving spotlighter, by contrast, has no obvious account to give. The presentist really does believe that everything is present and the growing blocker really does believe that reality is a block that grows over time, but the moving spotlighter doesn't *really* believe in a spotlight: that is just a metaphor. So what is the literal truth behind the metaphor? What is reality like that makes the metaphor apt? Out of all the times there are, why is one privileged, and in what way is it so? It is not obvious that the moving spotlighter has a good answer. Of course, the moving spotlighter could take it as primitive—there is just a primitive property of *being present* that some time has, end of story. But that is unsatisfying, and leaves us wondering if we even understand the moving spotlight theory: is there anything really underlying the metaphor?[16] So we have here a threat for the moving spotlight view, but not for presentism or the growing block, seemingly.

Table I.1 shows the score as it seems to be given by the preceding discussion. A tick means the view is doing well with respect to that issue, a cross that it is doing badly.

If the results in Table I. 1 are right, then there are issues on which presentism comes out the unique winner, issues on which the growing block comes out the unique winner, issues on which the growing block and the moving spotlight are tied as winners, and issues on which presentism and the growing block are tied as winners. But there is no issue on which the moving spotlight comes out as the unique winner. That suggests a dominance argument against the moving spotlight view: there is never a reason to prefer it over one of its two main rivals. It might be hard to choose between presentism and the growing block, since each does better than the other on different issues, but you should definitely reject the moving spotlight since it never does better than both its rivals and does worse than each on some issues.

[16] Cf. Trenton Merricks (2006, p104, fn.1), who says "I suspect there is no coherent story to be told about what, according to [the moving spotlight] view, being present amounts to."

Table I.1 Common problems for theories of time

	Presentism	Growing Block	Moving Spotlight
The epistemological problem	✓	✗	✗
McTaggart's Paradox	✓	✗	✗
Truthmaking	✗	✓	✓
Relations to non-present beings	✗	✓	✓
The open future	✗	✓	✗
What *is* presentness?	✓	✓	✗

The aim of this book is to develop a version of the moving spotlight view that fares better. There now follows an outline of the different chapters, giving a brief statement of what will be argued and how that ties into the issues discussed in this section.

Chapter 1: From A-Theory to Presentism? Part 1: Epistemology

In this chapter I argue that the presentist faces the epistemological problem just as much as the growing block and moving spotlight theorists. The orthodoxy in the literature—and what underlies the natural thought described in the previous section—is that the presentist can know that this time is the present because it is the only time there is. But, I will argue, that only means that the presentist is guaranteed to be *correct* when she claims that she is present, it does not guarantee that she *knows* this. To know that we are present, we need to be able to point to evidence that rules out the situation that we are non-present: and the presentist, growing blocker, and moving spotlighter alike need to be able to say what it is about their current evidence base that lets them know that this time is the present time. *Prima facie*, then, there is an epistemic puzzle for all A-Theorists.

I will look at a number of attempts to put the presentist back on a firmer footing, but I will argue that each of them fails. In essence, a recurring complaint will be that there are theses the presentist can appeal to that, if true, guarantee that she is present, but the question simply becomes: how can she *know* that they are true? And, I will argue, that just takes us back to the epistemic problem we started with. I will offer the presentist a solution, but it will turn out to be a solution that all A-Theorists can help themselves to. I will argue that the presentist should claim to know that presentism is true because it is the best theory for reasons concerning theory choice due to theoretical virtues, and that they know that they are present because this is a known consequence of the best theory. But I will also argue that the growing blocker and moving spotlighter can equally make this move, and hence can also know that this time is the objective present. Nonetheless, I will attempt to show that at least on certain ways of developing a non-presentist A-Theoretic metaphysics, there is both an epistemic and a metaphysical *cost* to those theories. It is not, however, inevitable that non-presentist A-Theories be saddled with these costs; the metaphysic defended in chapter four, I will argue, is immune from them.

Chapter 2: From A-Theory to Presentism? Part 2: McTaggart's Paradox

This chapter begins by examining, and dismissing, McTaggart's circularity and regress arguments that aim to rule out all versions of the A-Theory. I will propose an account of when circularity and regress are vicious—looking at examples of each from outwith the metaphysics of time—and I will argue that neither the circularity nor the regress that McTaggart identifies is vicious. I will go on to look at a modern attempt to rehabilitate McTaggart's argument by Nick J. J. Smith, and I will argue that this argument also fails. I will conclude that there is no McTaggart-esque argument that will refute all versions of the A-Theory. This discussion will draw heavily on the metaphysics of modality, using resources that have become familiar in that literature to advance this old debate in the metaphysics of time.

However, I will go on to argue that there is a McTaggart-esque argument that is a *prima facie* problem for *non-presentist* A-Theories. I will also propose a solution: the non-presentist *can* consistently avoid McTaggart's paradox. But, I will argue, they can only do so by giving up on the claim that if something was the case then it is the case in the past. This means that the non-presentist must abandon any claim to be solving the truthmaker problem for historical truths simply by believing in the existence of the past: that problem is only immediately solved on the assumption that the past *is* the way things were. This threatens to render the non-presentist ontology redundant, and this yields a challenge that the metaphysic to be defended must meet.

I will go on to look at Kit Fine's version of McTaggart's paradox, and argue against his claim that standard A-Theory does not fare well in responding to it. The discussion will bring out certain *constraints* that any standard A-Theory must meet if it is to be successful, but that is all. These constraints will guide us in our construction of the moving spotlight metaphysics in chapter four.

Chapter 3: On Giving an Ontological Account of Tense

In chapter two it was argued that the non-presentist has to reject the claim that how things were is how things are in the past. Does this mean that she cannot provide an ontological underpinning of claims about the past? It depends on what it means to give an ontological underpinning of some phenomenon. In this chapter I compare two meta-metaphysical approaches, each of which gives different answers to that. There is what I call the Quine–Lewis–Sider position, on which giving an ontological underpinning of tense is to say *what it is* in tenseless terms for a tensed claim to be true. If this is how one thinks of things, the moving spotlight view, it will be shown, is rather unattractive. I will advocate instead a truthmaker approach, whereby giving an ontological underpinning of tense is to say *what makes it the case* that the tensed truths are true. On this way of thinking of things, I will argue, the moving spotlight view is more attractive.

Chapter 4: The Moving Spotlight

Building on the meta-metaphysical view developed in chapter three, chapter four develops a moving spotlight metaphysic that provides truthmakers for each tensed truth, whilst avoiding the problems and meeting the constraints that have been brought out in the previous chapters. What is crucial in allowing the view to do the required truthmaking work while avoiding McTaggart's paradox is that the view shares with presentism the thesis that how things are *now* is how they are *simpliciter*. Where the view departs from presentism is not in admitting the reality of how things were, but rather in allowing that there are non-present entities that are nevertheless some way *now*. I also argue that in order to be an endurantist about how objects persist through time one has to be an A-Theorist, and in particular that one has to accept the thesis that how things are now is how things are simpliciter. In the course of giving this metaphysic, we also reveal a satisfying answer to what presentness consists in for the moving spotlighter.

Chapter 5: The Open Future

In this chapter I will first argue against branching ontologies on which the openness of the future consists in the reality of multiple future histories. I will then present an account of the open future where the distinction between the open future and fixed past can be secured without accepting that the past is real and the future unreal. The account is one I have defended previously with

Elizabeth Barnes.[17] It takes the openness of the future to be a matter of metaphysical unsettledness concerning what will happen. Following Barnes[18] and Barnes and Williams[19] I will argue that it is coherent to hold, and that there are good reasons to hold, that the world itself is unsettled in some respects—that is, that it can be unsettled whether something is the case and this not be a matter of our using imprecise language or concepts to describe or represent what is the case or a matter of our epistemic limitations. I will defend the claim that there are two distinct potential phenomena of metaphysical unsettledness: there can fail to be a fact of the matter concerning whether something is the case, and it can be indeterminate whether something is the case. The former is a matter of underdetermination in reality: there is no fact of the matter whether p when reality is not rich enough to speak to the issue. Indeterminacy concerning p arises, by contrast, when reality speaks to whether or not p is the case, but it is unsettled which state of affairs obtains. So when there is no fact of the matter whether p, both the truth and the falsity of p are ruled out; but when p is indeterminate, it is either true or false—neither is ruled out—but it is indeterminate which.

I will use this to characterize two ways in which the future might be open. There might be no fact of the matter concerning whether some future contingent is true, or it might be indeterminate whether it is true. I will argue that the former account sits well with the growing block view and the latter account with the moving spotlight. But I will argue that there is independent reason to prefer the indeterminacy account to the no fact of the matter account, and hence that the growing block view is at a disadvantage over its rivals.

Conclusion

After all that, Table I.1 should now look as shown in Table I.2.

Thus a dominance argument is made for the moving spotlight: it never does worse than a rival theory, and for each rival theory it does better than it in at least one respect.

Most of the changes to Table I.1 should be clear given what I have said that I am going to argue for. One that might not be obvious is the cross for the growing block view under "Truthmaking." This is tied to that view's failure to adequately deal with the open future (as will be argued in chapter five): there *should* be truthmakers for claims about the future, I will argue, because there *is* a way the future will be. It is simply that it is indeterminate *which* way the future

[17] Barnes and Cameron (2009, 2011). [18] Barnes (2010).
[19] Barnes and Williams (2011).

Table I.2 Common problems for theories of time—updated

	Presentism	Growing Block	Moving Spotlight
The epistemological problem	✓	✓	✓
McTaggart's Paradox	✓	✓	✓
Truthmaking	✗	✗	✓
Relations to non-present beings	✗	?	✓
The open future	✓	✗	✓
What *is* presentness?	✓	✓	✓

will be. But that requires that the truthmakers for claims about the future be indeterminate, not that there *aren't* any such truthmakers, *contra* the growing-block view.

The question mark for the growing block under "Relations to non-present beings" reflects something I won't go into in the chapters that follow, but I will say something briefly about now. I think the growing block view does all right on this issue, but arguably the moving spotlight view does better. It is not crazy to hold, as the growing blocker must, that we stand in no relations to future entities, in the way that it does seem to be a real cost to deny, as the presentist must, that we stand in relations to past entities. But nonetheless, I think it is plausible that we do stand in some relations to future entities, and so the moving spotlighter is at least at an advantage over the growing blocker here. I have friends who describe themselves as having, from the day their daughter was born, prayed for her future husband, that his life go a certain way. As it turned out, their daughter married someone younger than herself. *Prima facie*, then, there was a time when my friends were praying for someone who was yet to be born. Assuming, as I think is plausible, that one can only have a *de re* thought about something if it exists, this entails that the future being who was the object of their thoughts existed prior to his birth.

Now of course, the growing blocker can resist my friends' description of things: it would be easy to redescribe the case as them praying not *for* any particular person, but simply praying *that* their daughter gets married to some person or other whose life has gone a certain way. But this would be to *re*describe things; it is not what my friends *thought* they were doing—they thought they were praying *de re* for some particular person. They didn't know who it was, but they nonetheless thought that there was some person for whom they were praying. You can question whether prayer is worthwhile; you can object to their heteronormative assumption that their daughter will choose to get married to a man; but it seems to me wrong to tell them that we have discovered by doing some metaphysics that they are wrong about the content of their mental states. And my friends' case is not unusual, I don't think: we tend to have hopes, fears, beliefs, etc. about particular non-present things, both past and future. Other things being equal, at least, it would be better to accept a metaphysic—like the moving spotlight—that allows that we can stand in relations to, and *a fortiori* have propositional attitudes about, future entities.[20]

There is perhaps an obvious omission in the topics being discussed: the issue of how the moving spotlight metaphysic fits with special or general relativity. I will not be discussing that in the chapters that follow; other people are more qualified to speak to this issue than I am, and discussing the issues I want to discuss takes up enough room as it is. But I will say what would and would not worry me. Here are three potential objections that could be leveled against the moving spotlight metaphysic (call it "MS") defended in this book.

Objection 1: MS is logically (or a priori, or analytically) incompatible with any reasonable interpretation of our best current science.

Objection 2: MS proposes things that are not sanctioned by our best current science: that is, it posits features of reality for which there is no empirical evidence and which are of no predictive or explanatory use to our best scientific theorizing.

Objection 3: The package of our best interpretation of our best current science combined with a rival metaphysic to MS is overall more theoretically virtuous than the package of our best interpretation of our best current science combined with MS.

If *Objection 1* could be sustained, I would take that to be basically a refutation of the view—unless and until our scientific understanding changes to allow for this metaphysic. My credence that this can be shown, however, is low. Logical (or a priori, or analytic) incompatibility is a high bar, and I would be shocked if there were not at least a way of reasonably interpreting our current best science such that it is logically (etc.) consistent to accept both that interpretation of the science and the metaphysic defended here.

Objection 2 is almost certainly correct, but I do not consider it to be any objection at all. There is no point in engaging in metaphysics—indeed, there is no point in engaging in any descriptive discipline other than science—unless you think those disciplines can tell us things that science cannot. If you think

[20] Here I think the metaphysics of time contrasts with the metaphysics of modality. I doubt you would get anyone but a philosopher claiming to be able to think *de re* thoughts about a non-actual object.

metaphysics is pointless, this book is not for you; if you do not think that, you should not think it an objection that the metaphysic being defended tells us things about the world that no amount of empirical science will confirm.

Perhaps someone supporting objection 2 will have no objection to metaphysics telling us something science cannot confirm but will think there is a particular problem with it telling us something about *time* that science cannot confirm. After all, the thought goes, time is one of the things that science aims to tell us about, so we should expect truths about time to be amenable to scientific discovery. Perhaps metaphysics can tell us that there are moral properties, or possible worlds, or that people could survive teletransportation, but it cannot tell us that something within the domain of scientific enquiry has some features that science is blind to. But again, I do not think we should accept this constraint on what metaphysics can tell us. If you hold that metaphysics can tell us things that science cannot, I think you should also allow that it might reveal to us more facts about the kinds of things that science tells us some things about. Science tells us some things about the nature of ordinary objects—such as that they are mostly empty space, that each one exerts a gravitational pull on each other one, etc.; but I think we should still allow that metaphysics can help us with questions about the nature of ordinary objects that science can never answer—such as, for example, whether they are bundles of properties, or consist of properties and a bare substratum, or something else. I do not see why the nature of time cannot be the same, with science revealing some facts about its nature and metaphysics revealing others. Of course, this is not going to convince anyone who thinks that there is simply no role for metaphysics to tell us things about the world, but that is not my aim.

Objection 3 is where the action is, to my mind. It would be an objection to—although not a refutation of—a metaphysical view if an overall theory of the world that includes our best interpretation of our current best science and a different metaphysical view is more virtuous than an overall theory of the world that includes that science and that metaphysic.

I say it would not be a refutation, because it might be the case that there is a less good interpretation of the science that nonetheless, when combined with that metaphysic, makes for the overall best worldview. It is not clear to me that science trumps other disciplines such that we can be certain that what is intrinsically the best interpretation of our current best science will thereby be part of the best overall theory of the world. Perhaps a slightly less virtuous interpretation of the science allows for a massively more virtuous metaphysic, or a massively more virtuous account of moral reality, etc., such that overall we should accept a worldview that includes this intrinsically less virtuous scientific view.[21] Nonetheless, it would be at least a *pro tanto* objection to the moving spotlight metaphysic if a rival metaphysic was a more virtuous fit with our best science. But in order to know whether this is the case, we of course need to know what theoretical benefits are obtained by believing the moving spotlight metaphysic. This is what I will attempt to show in this book. So while I am not explicitly addressing the question of the moving spotlight's fit with special or general relativity, you can view my book as providing some of the data that are necessary in assessing any such objection to the moving spotlight.

[21] There could be an objection that sat between objections 1 and 3: that MS is logically (or a priori, or analytically) incompatible not with *any* reasonable interpretation of our best science, but with the best such interpretation. But my reply to objection 3 holds here as well: such-and-such being the best interpretation of the science is compatible with it not being part of the best overall theory of the world. Perhaps, e.g., a *slightly* less good interpretation of the science allows for a *massively* better metaphysical theory of the world, resulting in this being a better overall theory of reality.

1

From A-Theory to Presentism?

Part 1: Epistemology

If, as the moving spotlighter thinks, there are many times, one of which is objectively present, how could we know *which* of the many times is the present? Perhaps we are stuck in the distant past or far future. Presentists attempt to secure knowledge of our presentness by saying that there is only one time, so of course it is that time that is present; Peter Forrest attempts to secure knowledge of our presentness by saying that non-present beings lack consciousness, so that only present beings can entertain the thought that they are present. I begin (§1.1–1.2) by arguing that neither approach solves the problem, for they merely shift the challenge to how we know the metaphysics in question. And so every A-Theorist faces the epistemic puzzle. In §1.3 I argue that any A-Theorist can solve the problem if she appeals to a liberal externalist epistemology; I argue that even given a view on which there are many past and future beings who think that they are present, *our* belief that we are present is still *safe*, and our belief-forming process *reliable*. In §1.4 I return to Forrest's dead past hypothesis and argue that there are reasons to accept something like it after all—just not that it solves the epistemic puzzle. This is an issue we will return to in chapter four.

1.1 A Skeptical Puzzle for Non-presentist A-Theories

In the first two chapters we will look at arguments that purport to take you from the truth of the A-Theory to presentism: that is, they aim to show that to avoid some potential problem, the A-Theorist must limit her ontology to present entities. There are two potential problems for the non-presentist A-Theorist that we will look at: one epistemological and one metaphysical. In each case I will argue that the non-presentist A-Theorist can avoid the problem, but each discussion will reveal certain constraints that must be met by any acceptable non-presentist A-Theory, and these will guide the subsequent discussion in the book. We will start with the epistemic objection.

Non-presentist A-Theories like the growing block and moving spotlight views have been thought to face an embarrassing epistemic problem: if they are true, we cannot know whether or not we are present.[1] After all, if presentism is false then there are many non-present people who believe they are present. Their pool of evidence does not appear to be any worse than ours, so how can *we* know that we are present?

There is Caesar, for example, in the past, thinking to himself "I am now crossing the Rubicon." He's wrong![2] Caesar's crossing the Rubicon happened in the past; he's *not* now crossing the Rubicon. But here is me thinking "I'm now writing this chapter." Why should I think I'm any better off than Caesar? Perhaps someone is to me as I am to Caesar, looking back on my thought and thinking "There's Ross thinking he's present and not knowing he's past."

The B-Theorists face no such problem. According to them, the truth-condition of Caesar's thought is that it occurs at the same time as his crossing the Rubicon, whereas the truth-condition of my thought is that it occurs at the same time as my writing the chapter. We are each in a position to know that those conditions are met. And orthodoxy has it that the presentist also is immune to the problem. According to the presentist, Caesar is not thinking that he is now crossing the Rubicon, because Caesar does not exist. The only things thinking thoughts of the form "It is now p" are present beings, because *everything* is a present being. If presentism is true, you are *guaranteed* to get it right if you think you are present. If *everything* is present, so the thought goes, then knowing that you are present is no harder than knowing that you exist; and that is easy to know. The problem only arises, so orthodoxy says, when you combine the objectivity of the present with the existence of non-present beings: for then I need to know whether *this* time is the objective present in order to know whether claims of the form "It is now p" are true, and yet being objectively present does not appear to change the evidence available to the beings who inhabit that time.

Of course, the non-presentist A-Theorist could deny this last assertion and claim a way out of the puzzle that way. Peter Forrest,[3] for example, believes in past entities as well as present ones, but thinks that only present entities are conscious. The past is populated by people lacking in consciousness. So while Caesar exists, and is crossing the Rubicon in the past, it is a Caesar lacking in

[1] See Bourne (2002, 2006), Braddon-Mitchell (2004), Heathwood (2005), and Merricks (2006).

[2] I assume that "It is now the case that p" has the truth-condition that p is the case at the *present* time, so that being mistaken about what is happening now goes hand-in-hand with being mistaken about what is present. Nothing hangs on this, however: we could instead simply focus on Caesar's belief that his crossing the Rubicon is a present event. It might be rarer to explicitly think that things are present than to think that they are happening now, but as long as it happens sometimes—and it does!—then we can generate the puzzle, even if the assumption is false.

[3] Forrest (2004).

consciousness that is crossing the Rubicon. Of course, *when* that event was present, Caesar was conscious, but as soon as it became past, he was no longer so. So being at the objectively present time *does* change the evidence available to you: and since you can detect that you are conscious you can thereby, thinks Forrest, detect that you are present.

In fact, I agree with Forrest that Caesar is no longer conscious: indeed, as we shall see in chapter four, I think that Caesar is *very* different from how he was when he was present. But as I will argue below, this doesn't in fact solve the epistemic problem. My reasons for thinking that Caesar has changed in these respects are metaphysical, not epistemic, and my solution to the epistemic puzzle will not rely on that. So for now, and until the final section of this chapter, let us simply assume that Caesar is having experiences of a similar character to our own and that he believes that he is having those experiences in the present. Even on this assumption, I will argue, the epistemic puzzle can be solved. But before I try to solve the puzzle, I want to make it as hard as I can; in the next section, I will argue that the puzzle faces *every* A-Theorist, not only the non-presentist.

1.2 Why, Given Strict Standards for Knowledge, the Problem Is just as Pressing for the Presentist

If the past is real—as the moving spotlighter and growing blocker think—then, *prima facie*, how things were is how the past is. So since Caesar certainly *was* having certain experiences and thinking that they are present, Caesar *is* in the past having those experiences and thinking that they are present. In that case, it seems that there is no difference in character between his experiences and our own that would give us evidence that we are present. Perhaps then, in order to explain our knowledge that we are present, we need to deny that Caesar exists. The orthodoxy in the literature is that the presentist does not face the skeptical puzzle: that if there simply *are* no non-present beings, then there is no puzzle in knowing that you are present.

The presentists that have pushed this epistemic objection against non-presentist A-Theories have not tended to make explicit their views on the nature of knowledge or justification, but one thing I hope to bring out is that this makes a big difference to how damning the epistemic objection is. In this section, I will assume we are operating with strict internalist requirements on knowledge. In particular, I will assume in this section that if you know that p then you must have access to the basis for your knowledge that p: there must be something about your evidence base, accessible to you, that rules out it being the case that not-p.

If these are the standards required for knowledge then, I will argue, there is a problem for *all* A-Theories: even if presentism is true, it is hard to see how we could know that we are present. In the subsequent section I will assume looser standards for knowledge, such that an externalist might be happy with them; if *those* are the standards required for knowledge then, I will argue, there is no problem for *any* A-Theorist.

For now, then, let us assume that in order to know that we are present there has to be some evidence accessible to us that would form a basis for this knowledge. It is clear to see why, given such standards, the non-presentist A-Theorist faces a puzzle: it looks like there is *no* difference in character between the experience I am having now and the experience Caesar had in the past, and no relevant difference in mental state. Hence, there is nothing I can detect in my evidence basis that rules out my being in the same position as Caesar: that is, stuck in the past. Hence, by internalist standards, I cannot know that I am present, even if I happen to be so, if the moving spotlight or growing block is true. But if knowledge is so hard to obtain, I think we are in trouble even if presentism is true. There are some things the presentist can say to justify her claim to knowledge of her own presentness even by internalist standards, but they are all such that if they are successful for the presentist then they are equally available as solutions for the non-presentist A-Theorist, and so there is no epistemic advantage to presentism.

Presentists who have posed the skeptical puzzle to non-presentist A-Theorists have said very little as to why the presentist is supposed to avoid the problem. Heathwood simply states that presentism is "immune to scepticism about the present," but offers no argument.[4] And Bourne says simply

> It is clear that presentism...solves the [skeptical] problem, not due to any distinct phenomenological experience, but simply because if we only initially invoke the existence of our present time as the one real time, we could not help being [objectively present], since *ex hypothesi* it is not possible for us to be anywhere else: *I am, therefore I am present.*[5]

And Braddon-Mitchell agrees

> [H]ow [do] we know that the current time really is the present? If you are a presentist then [there are] no epistemic problems. For according to the presentist all that exists is the present, so the fact that we know we exist guarantees that we are in the present. The presentist has an objectively characterized conception of the present, but it is one we have simple epistemic access to.[6]

[4] Heathwood (2005, p250). [5] Bourne (2006, p24).
[6] Braddon-Mitchell (2004, p199).

But I think that, on examination, it will be revealed that the puzzle is just as pressing for the presentist. To begin, notice that it cannot simply be (as Bourne and Braddon-Mitchell seem to think) that the presentist avoids the problem purely because she does not believe in the existence of people thinking they are present and getting it wrong. Thinking that there are many brains in vats undergoing simulated experiences of being in a world like ours might make *vivid* the skeptical problem of how you know you are not such a brain in a vat, but it is merely a rhetorical device. What is crucial to the case for external world skepticism is not that there *are* many brains in vats in similar epistemic scenarios to your own, but simply that *were* you a brain in a vat you *would* be in a similar epistemic scenario to your current one; that is sufficient to make it puzzling how you can know you are not a brain in a vat.

Similarly, the *existence* of non-present beings is irrelevant to whether you can know that you are present. What matters is whether your evidence permits you to rule out the scenario that you are not present. Thinking about poor Caesar makes it *vivid* that you cannot, since it seems evident that you are not better off epistemically than he is, if he exists. But simply refusing to believe in Caesar and his non-present companions is not sufficient for solving the problem. Everyone might be present and yet have insufficient evidence to rule out their being non-present, just as everyone might be enjoying the external world and yet have insufficient evidence to rule out their being a brain in a vat.

So if presentism avoids the puzzle it must be due to some other feature of it, not simply to the presentist's insistence that nobody wrongly thinks they are present. If that is true, it ensures that I am always *correct* when I think I am present, but it does not ensure that I *know* it.

Nor is it the *modal* status of presentism that makes the difference. The presentist will likely think it *metaphysically necessary* that everything is present, and so it is not merely that no one gets it wrong when they think they are present, but that no one *could* get it wrong. By contrast, most people will think that I *could* have been a brain in a vat, even if no one in fact is. But this is not a relevant disanalogy. What matters is whether I can rule out the scenario in which I am non-present, or in which I am a brain in a vat, and it does not aid my doing so for the scenario in question to be metaphysically impossible.[7] Water could not be XYZ, but this plays no role in my ruling out the epistemically possible world in

[7] Descartes, in his *Meditations*, correctly, does not take himself to have solved the "evil demon" skeptical problem simply by saying that there is a necessarily existing God who would not deceive us. We need to *know* that there is such a God who would not deceive us to rule out the demon scenario. Similarly, we need to *know* that there could not be non-present things if we are to know that we cannot wrongly think we are present. But any pretensions to such knowledge *presuppose* that we can

which it is so. The mere epistemic possibility of my being non-present is enough to generate the puzzle. Whether this is a genuine metaphysical possibility is neither here nor there. The necessity of presentism entails the *safety* of my belief that I am present, and the *reliability* of any process that results in your coming to believe that you are present, but it does so by making my belief necessary;[8] and it simply cannot be a sufficient condition for knowledge of necessary truths that my belief in them be safe, or that there be a reliable process that results in such a belief, for that would make such knowledge too easy to obtain. So it does not seem that taking presentism to be necessary rather than merely true helps explain how the presentist knows that she is present.

It is tempting for the presentist simply to reason thus: "I know I exist. I know that everything that exists is present. So I know that I am present." But this will not do as it merely postpones the problem, for the puzzle now becomes: *how* can they know that everything is present? What in their evidence lets them rule out the scenario in which there are non-present things? If they cannot rule out the scenario that *they* are non-present, then they cannot rule out the scenario that there are *some* things that are non-present; and so they cannot rely on the claim that they know that everything that exists is present when arguing for the claim that they know that *they* are present.

Suppose I am worried about the problem of other minds, and in particular I am wondering if I have any reason to believe that Sara has an inner mental life. A panpsychist tries to convince me not to worry: Sara has a mental life because *everything* does. Surely, that is not at all helpful. Insofar as I am worried about whether my evidence supports my believing that Sara has a mind, I should be at least as worried about whether it supports my believing that *everything* does. The truth of panpsychism presupposes that Sara has a mind: if my evidence is compatible with Sara having no mind then it is compatible with something having no mind, *a fortiori* it is compatible with the falsity of panpsychism. Likewise, the truth of presentism presupposes that I am present: if my evidence is compatible with me not being present then it is compatible with something not being present, *a fortiori* it is compatible with the falsity of presentism. So either my evidence is compatible with me being non-present—in which case I cannot know that presentism is true—or it is not compatible with me being non-present,

already rule out that we are non-present, so we are not going to make headway on the problem this way.

[8] Or at least, if it is contingent that I am present that is only because I am a contingent being. The necessity of presentism guarantees the necessary truth of the conditional: *if* I exist, then I am present. But the point still stands: that cannot be sufficient for my belief in this conditional to amount to knowledge, for that makes knowledge of this deep truth of metaphysics (assuming it is true) too easy.

in which case I do not have to go via the truth of presentism to justify my belief that I am present, I just have to say what about my evidence rules out my being non-present. In neither case, then, is the question as to how I can know that I am present helped by answering "Because everything is." It is dialectically inappropriate for the presentist to assume knowledge of their theory of time when the question is how we can know this claim about our position in time (i.e. that we are present). Our warrant for believing that theory of time is *threatened* if we cannot know the claim about our position in time, so it would be dialectically inappropriate simply to assume that we know the theory to be true when defending our claim to know this fact about our position in time.

Perhaps the presentist will claim that it is a priori that everything is present. But again, this is not a solution, it is just postponing the issue: the question becomes *how* can one know this a priori? As puzzling as it is how your experiences can let you rule out the scenario that you are non-present, so is it puzzling how the light of pure reason lets you rule this out. It is not, after all, a conceptual necessity that everything is present: it is implausible that the eternalist is entangled in some conceptual confusion in positing non-present things. The presentist can understand the eternalist's view that there are non-present beings—she just thinks it is false. And just as it is implausible that eternalism is conceptually incoherent, so is it implausible that the eternalist's claims are analytically false. If it was a matter of the *meaning* of "present" that everything is present then knowledge of one's presentness could be secured in much the same way as it is if the B-Theory is true, so the presentist who thinks that it is analytic that everything is present faces no epistemic problem. But this is a bad version of presentism and it should be rejected: it is implausible that non-presentists are saying something analytically false, and to suggest that they are simply betrays a misunderstanding of English. The moving spotlighter (e.g.) and the presentist are engaged in a substantive debate about the nature of reality: they are not talking past one another or having a debate about the meaning of "present" in English. Of course, one can use one's terms as one wishes, and anyone is free to mean by "present" something that secures the analyticity of "everything is present." But to do so is just to change the subject. The challenge is to say how we can know that we are *present*: you cannot decide to mean something else by "present" and show how you can know the proposition expressed by the sentence "I am present" given this new meaning and still claim to be addressing the same problem.

It is neither conceptually necessary nor analytically true that everything is present, and so to claim that it is a priori (for me) that I am present does not solve the puzzle of *how* it is that I can know that I am present. Of course, the presentist could always just claim that we have a rational insight into the truth of "everything is present," and refuse the demand for further explanation. But if that move is acceptable for the presentist, why can't the A-Theoretic non-presentist simply claim that we have a rational insight into the fact that it is *this* time that is present, and refuse any further demand for explanation? To plead rational insight is to refuse to give an explanatory account of the knowledge that we are present and just to claim that we have it via some mysterious process: and the eternalist and presentist alike can make that move.

One suggestion I have heard made on a number of occasions in conversation is that "I am present" has a "cogito-like status" for the presentist but not the A-Theoretic non-presentist, and this justifies the presentist in believing that she is present in a way that is unavailable to her non-presentist A-Theorist rivals. But what does it mean to say something has a "cogito-like status"? I can think of a few suggestions, but on none of them does it turn out to be both true and helpful that "I am present" has a cogito-like status for the presentist but not for the non-presentist.

To say that "I am present" is cogito-like could be simply to claim that it cannot be falsely uttered. Then it is true that it is cogito-like given presentism but not given non-presentism. But as I have already argued: this is only enough to guarantee that we are always correct when we claim that we are present, not that we know it to be so. Or perhaps to say that it has a cogito-like status is to say that it cannot be coherently doubted. But then it is simply false that "I am present" has a cogito-like status: this can be coherently doubted because presentism can be coherently doubted. The non-presentist A-Theories are not unintelligible, and supposing that I am not present does not have the self-defeating aspect that supposing I do not exist does. What is true is that "Presentism is true and I am not present" cannot coherently be believed. But this does not help secure our knowledge that we are present unless we already know that presentism *is* true, which just brings us back to the problems above. Perhaps to say that something is cogito-like is to say that it follows from something for which we have direct evidence, just as "I exist" follows from "I think," the latter of which we have direct evidence for simply from introspecting. Well, of course the presentist thinks it follows from my thinking that I am present. But again, this is not enough to secure *knowledge* that we are present: she has to *know* that it so follows. The epistemic puzzle has not been solved, merely shifted.

And now we can see why Forrest's proposal does not actually help with the epistemic puzzle for the non-presentist A-Theorist. Forrest also thinks that "I am present" is cogito-like in this last sense: he thinks that my being present follows from my being conscious, the latter of which I have direct evidence of via introspection. It should be clear now, given what I have said about presentism, that this does not solve the epistemic puzzle. Forrest's view entails, just like presentism, that everyone who thinks that they are present is correct. But just as with presentism, this does not mean that if Forrest's view is right then we can know that we are present: we'd have to *know* that only conscious beings are present. Forrest, like the presentist, has not accounted for our knowledge that we are present, he has merely shifted the epistemic puzzle to how we can know that the metaphysic of time in question is true. If it is an epistemic possibility that I am not present, then (since I know for sure that I am conscious) it is an epistemic possibility that there is a conscious non-present being, which is just to say that it is an epistemic possibility that Forrest's metaphysic is false. Coming to know that it is true would require ruling out the epistemic possibility of there being a conscious non-present being, which would mean ruling out the epistemic possibility of my being non-present, which *just is* the epistemic puzzle in question. Forrest's metaphysic might do just as well as the presentist's concerning the epistemic puzzle; but the arguments of this section are intended to show that the presentist does not do as well as she thinks.

Perhaps the presentist should be less Cartesian and more Moorean, and claim not that "I am present" has a cogito-like status but that it has the status of a Moorean truth: a claim which is on a stronger epistemic footing than at least one of the premises in any valid argument for the conclusion that you might, for all you know, be in a skeptical scenario in which that Moorean truth is false. Perhaps the claim that I am present does indeed have (for me) the status of a Moorean truth. But why should it be of any particular advantage to the *presentist* should this be so? Wouldn't this simply give the non-presentist A-Theorist a way of resisting the epistemological puzzle that we are concerned with? "Ingenious argument, Madame Presentist, and I cannot pinpoint where it goes wrong; but I know that *something* does, since your conclusion is that I might be in the skeptical scenario in which I think I am present but am not. I can dismiss that argument because my belief that I am present is on a better epistemic footing than at least one of the premises of your argument." Replace "presentist" with "external world skeptic" and "I am present" with "I have hands" and we have here the Moorean response to the argument that you do not know that you have hands because you might be in the brain-in-a-vat scenario, in which you think you have hands but do not. So if the presentist knows that she is present because "I am present" is a Moorean truth (for her), then the non-presentist A-Theorist can resist the presentist's epistemological objection for the same reason.

Perhaps the presentist will agree that everyone—presentist and non-presentist alike—should take "I am present" to be a Moorean truth, but argue that this gives us reason to accept presentism, and hence that non-presentist A-Theorists who justify their claims to presentness by claiming it to be a Moorean truth that they are present are in a dialectically unstable position. Why think that claiming it to be a Moorean truth that you are present should lead you to accept presentism? Perhaps the thought is that we should not accept metaphysics that undermine our knowledge of those claims we profess to be Moorean truths. Well, that may be so: but whether or not non-presentist A-Theories *do* undermine our knowledge of our presentness is precisely what is up for debate, so the presentist cannot assume this without begging the question. Perhaps the claim is that we should not accept a metaphysic which renders unsafe our belief in claims we profess to be Moorean truths. Again, that may be so: but as I will argue later, non-presentist A-Theories do *not* render unsafe our belief in the claim that we are present. So I do not think attributing the status of Moorean truth to "I am present" is going to put the presentist at an epistemic advantage any more than claiming it to be cogito-like did.

I said earlier in this section that the modal status of presentism is not sufficient to solve the puzzle, but perhaps the presentist will claim that it is not merely necessary that what exists is present, but that *what it is* to exist is to be present: so there is a hyperintensional connection between being present and existing, not merely necessary equivalence. To the question "How do you know you are present?," this presentist's response will be "Because I know that I exist, and that's just *what it is* to be present." This move, of course, is not open to non-presentist A-Theorists: they certainly do not think that what it is to be present is to exist, given that they believe in non-present entities.

But this still does not help: it is not enough for it simply to be true that what it is to exist is to be present. As just stated, that guarantees that anyone who thinks that they are present is *correct*, but it does not follow that they *know* it. To secure the knowledge that you are present, you would have to *know* that what it is to exist is to be present. So how could the presentist come to know that? Some *what it is* claims can be known as a result of conceptual analysis. Anyone who competently grasps the concept *vixen*, for example, is in a position to know that what it is to be a vixen is to be a female fox. But that is definitely not what is going on in this case: as already mentioned, it is implausible that the eternalist, in admitting the existence of non-present things, is engaged in some kind of conceptual confusion. Presentism is not true as a matter of conceptual necessity even if it is metaphysically necessary. Other *what it is* claims are known as a result of empirical investigation: the empirical evidence that gives me reason to believe that water is H_2O also gives me reason to believe that *what it is* to be water is to be H_2O. But what evidence does the presentist have available to them that gives them reason to believe that what it is to be present is to exist? In the water case, we know a priori that *whatever* the chemical structure of water, *what it is* to be water is to be of *that* chemical structure, and then we discover empirically that that chemical structure is H_2O. Perhaps the presentist will claim to know a priori that *if* everything is present then *what it is* to exist is to be present. But the puzzle is exactly how they can come to know the antecedent: it is puzzling *how* the presentist can know they are present (let alone that *everything* is). Perhaps the presentist will simply claim to have direct rational insight into the claim that *what it is* to exist is to be present. But as above, if we are allowed to start pleading rational insight, I do not see why the non-presentist cannot simply claim rational insight into *this* time being the present one.

Perhaps this presentist will try a different tactic. Perhaps she will claim that she does not need to *know* that what it is to exist just is to be present in order to know that she is present: it is enough that that is the case. The thought is: she knows that she exists, for cogito reasons; so she stands in the knowledge relation to the proposition that she exists; but since to exist *just is* to be present the proposition that she exists *just is* the proposition that she is present; and, as we have already established, she stands in the knowledge relation to that proposition; so she knows that she is present after all.

This is unsatisfying. Suppose we accept the legitimacy of the following chain of reasoning: (i) Proposition p has A as a constituent; (ii) A just is B; (iii) therefore, the proposition q that is obtained by replacing B for A in p just is the proposition p. Then of course if what it is to be present is to exist then the proposition that I exist just is the proposition that I am present, and so any relation I stand in to one I stand in to the "other" (for it is not really "other," but the very same proposition). *A fortiori*, any epistemic relation I stand in to the proposition that I exist—such as *knowing that*—I also stand in to the proposition that I am present. But if propositions are individuated thus, then it simply has to be the case that there is more to my epistemic state than what propositions I stand in the *knowing that* relation to. There has to be *some* sense in which people used to not know that water is H_2O, even though they knew that water is water, and water *just is* H_2O. Another example: suppose as a matter of fact that Julie is going to win the lottery, but that (as we would normally say, in any case) I do not know this. I can introduce "Sally" as a descriptive name for whoever will win the lottery, and I thereby know that Sally will win. But there has to be *some* sense in which I do not know that Julie is going to win, even though I know that Sally will win, and Sally *just is* Julie. One thought is that while I do indeed know the proposition that

Julie will win, and that prior to chemical investigation people did indeed know that water is H_2O, we did not know those propositions under those guises. And maybe there are other things one could say. But on any plausible account, there is some epistemic good that is lacking with respect to me and the fact that Julie will win and with respect to our early drinkers and the fact that water is H_2O. We gain *some* epistemic good when we are able to identify Julie as the person who won and H_2O as the chemical composition of water. Maybe it is that we stand in the knowledge relation to a new proposition; maybe it is that we bear the knowledge relation to a proposition we already bore it to but under a new guise; maybe it is something else. Whatever it is, it is an epistemic virtue to have it. And whatever that epistemic virtue is, we must be in a position to have it with respect to the fact that we are present. It is not enough for the presentist to convince us that we know that we are present in the same sense that prior to chemical investigation people knew that water was H_2O. If the presentist thinks that the proposition that she is present just is the proposition that she exists, then we should simply reformulate the puzzle: rather than asking how she can stand in the knowledge relation to that one proposition, we should instead ask how she can know that proposition under the guise of *the proposition that she is present*. The mere truth of the *what it is* claim, together with her knowledge of that proposition under the guise *the proposition that she exists*, will not settle that she also knows the proposition under the required guise. If she *knew* the *what it is* claim, then that would perhaps be sufficient: but this just brings us straight back to the problems mentioned above.

I think the best thing to say, if you are a presentist who thinks that what it is to exist is to be present, is that this *what it is* claim is justified simply because your best overall theory of the world says that it is true. Thus you give a pragmatic reason for believing the *what it is* claim: the principles of theory choice select a theory according to which it is true. I have no problem with the presentist pleading this kind of pragmatic justification for the claim; but we are now departing from the internalist assumptions at play in this section, and as I will argue in the next section, if this is a good route to knowledge, the non-presentist A-Theorist has a perfectly acceptable response to the skeptical puzzle.

Let me sum up the moral of section 1.2. Presentism entails that we are present. If presentism is true then we cannot wrongly think that we are present. But this does not secure our knowledge that we are present, unless we can come to *know* that presentism is true. And, so far, this looks like no easier an epistemic problem than the one we started with: knowing that *everything* is present is, if anything, harder than knowing that *you* are.

A comparison may help. Theodore Sider argues against the neo-Fregeans' claim that their theory dissolves any epistemic problem about how we can come to know about a realm of abstracta like numbers.[9] He says: "In effect, what neoFregeans are trying to do is argue for an underlying metaontology [namely, quantifier variance] ... that guarantees the success of their stipulation of Hume's Principle. They hope thereby to dispel doubts about mathematics. But in order to dispel all doubts, it is not enough that the underlying metaontology be *true*. It must itself be epistemically secure.... [But] the rejection of ontological realism in favor of quantifier variance is, if anything, less epistemically secure than the mathematical knowledge it is supposed to ground.... Far from providing a detour around substantive fundamental metaphysics, neoFregeanism is itself a piece of substantive fundamental metaphysics."[10] Sider's point, and I agree, is that it doesn't make the epistemology of mathematics easy if the existence of numbers follows trivially from a theory T, if it is a hard epistemological question as to whether T itself is true. The point, of course, has nothing to do with neo-Fregeanism or mathematics: in general, adopting a theory T that entails p only eases epistemological qualms concerning how we could come to know that p if T is on a firmer epistemic footing than p is independently of T. For if it is not, we have not solved the epistemic problem but merely shifted it, for now the question is simply: how do we know that T is true? And just as it is for neo-Fregeanism and truths of mathematics, so it is for presentism and the claim that we are present. Presentism is a substantive metaphysical theory, and the epistemology of metaphysics of time is difficult. Taking Sider's lesson on board, then: we do not resolve the epistemological worry about how we can know we are present simply by noting that presentism entails that we are. It is not enough that the underlying metaphysic be *true*. It must itself be epistemically secure. The rejection of eternalism in favor of presentism is, if anything, less epistemically secure than the knowledge of our presentness it is supposed to ground.[11]

Before we continue, this may be a good point to consider a rival approach that I think *would* easily solve the epistemic puzzle for the presentist and non-presentist A-Theorist alike, were it acceptable. The view is that presentness is detected in our experiences *directly*: that just as I simply see something's being red, so I simply see something's being present. Compare this view to Forrest's. Forrest is offering what we might call a reductive explanation of our knowledge of presentness: he pinpoints some *other* feature of our experiences that we can detect (in Forrest's case: the conscious nature of our experiences) and argues that

[9] See Wright (1983), Hale and Wright (2001).
[10] Sider (2007a, §3.1).
[11] Ironically, Sider is one of the people who simply assume that presentism is immune to the epistemological problem (Sider 2013a, p261).

that feature is only to be had if what is experienced is present. The reason this does not solve the epistemic puzzle, I argue, is that this only shifts the problem to how we can know that only present things have experiences of that kind. That "linking principle" between a feature that we can detect by introspection and the presentness of those experiences is a theoretical claim, on at least as bad an epistemic footing as the claim that we are present.

Contrast a non-reductivist who thinks that presentness *itself* is experienced: that when we experience a present event, part of our experience of the event is *that* it is present. Just as we can experience properties of something like its size, color, shape, etc., so can we experience its presentness, or so says the non-reductivist. If non-reductivism is true, then Caesar is not experiencing the presentness of his crossing the Rubicon, because it is not present; whereas my writing this chapter *is* present, and that is one of the properties of the event that I am experiencing the event as having. So for example, here is Brad Skow, on behalf of the moving spotlighter:

> Of all the experiences I will ever have, some of them are special. Those are the ones that I am having NOW. All those others are ghostly and insubstantial. But which experiences have this special feature keeps changing. The moving spotlight theory explains this feature of experience: the vivid experiences are the ones the spotlight shines upon. As the spotlight moves, there are changes in which experiences are vivid.[12]

Note that Skow is not here talking about the experienc*ed* but about the experience: what we have access to is some of our experiences being special, and this is the *reason* for thinking there is a corresponding specialness in the reality that is being experienced. As Skow sees it, certain of our experiences are specially vivid, and the objective present of the object of such experiences is postulated to *explain* that vividness in experience. And plausibly, the vividness of the experience *just is* the perception of the presentness of the object of experience, just as the redness of an experience is the perception of the redness of the object of experience.

I think if we accept this non-reductivist view then we do indeed solve the epistemic puzzle. The difference between this view and Forrest's is that the presentness of events is not being *inferred*—rather, I simply see their presentness. When presentness is inferred from some other feature—such as consciousness for Forrest, or existence for the presentist—then the question is simply shifted to how we know that the inference is a good one: how do we know that the metaphysical theory that says that all conscious (or existing) things are present is true? No such question arises if I simply *see* an event's presentness: then I have

[12] Skow (2009, §4).

as much evidence for the event's being present as I do for the fire truck's being red: in both cases, the evidence is that the thing having that feature is a constituent of my experience.

So the non-reductivist has a solution to the epistemic puzzle; but I do not think we should accept this solution, because I do not think that I *do* experience the presentness of the objects of my experience. Reflecting on the nature of my own experiences, I can see no reason for thinking that things being present is a way I experience things to be in the same way I experience things being, for example, red. Following L. A. Paul,[13] let's distinguish experiencing something as present from experiencing it *as of* present. To experience something as of present is to experience something in such a way that you thereby *take it* to be present, but the property of presentness need not itself be a constituent of the experience. It seems clear to me that my experiences are of things as of present: every experience I have, I take it to be a present experience. But the explanation of that could simply be to do with my psychological makeup (as Paul believes): there is no reason to think this feature of my experiences is to be explained by some feature of the events so experienced. And I have no reason for thinking that Caesar's experiences—if such there be—lack this feature, since it is perfectly consistent that some past events are nonetheless experienced *as of* present. To solve the puzzle I need to experience things as present: the presentness of things must itself be a feature of my experience of them. But why would I think I experience that feature of things, given that I have no *contrasting* experiences to compare to? I think I experience the redness of the fire truck, and part of my reason for thinking that I am experiencing a feature *of* the fire truck is that I can contrast the nature of this experience to my experiences of non-red things. I cannot simply be psychologically programmed to experience things *as of* red, since I have plenty of experiences that *do not* have that nature. The best explanation for why this experience has that nature when others do not is that I am experiencing a feature of the object of that experience that is lacked by the objects of those other experiences. But when it comes to presentness, I do not have a feature of some experiences that is lacked by others such that I can point to it and say that the reason some experiences have that feature is because I am detecting the presentness of the objects of experience. If I was experiencing, still, my fifth birthday party, but the experience was, as Skow says, "ghostly and insubstantial," then I could contrast this with my "vivid" experience of writing this chapter, and explain the vividness of the latter as the experience of the presentness of that event. But I am not experiencing my fifth birthday party. I do not appear to have

[13] Paul (2010).

any ghostly and insubstantial experiences, and none of my experiences appear peculiarly vivid. Whatever feature of my experiences I point to and claim it to be the experience of presentness, *all* my experiences have that feature. In which case, I have no reason to think that it is a feature that the properties of the things experienced are responsible for, as opposed to a feature that my psychological makeup is responsible for; I have no reason for thinking that I am experiencing things as present rather than merely experiencing them as of present. So I see no reason to accept the non-reductivist's story.

1.3 Why, Given Less Strict Standards for Knowledge, There Is a Solution for Presentist and Non-presentist Alike

Presentism says that everything is present, *a fortiori* we are present. We have been looking at how the presentist can claim to know the former, in order to thereby know the latter. Progress has not been made, because coming to know that everything is present looks to be just as difficult as the original puzzle: what about our (a priori and a posteriori) evidential base lets us rule out that there are non-present things?

Of course, we have been assuming some high standards for knowledge. We have been demanding that the presentist point to something in her evidence base that is accessible to her and which indicates that she is present. Perhaps it is no surprise that on such strict standards, knowledge of a deep metaphysical fact is unobtainable. Let us, then, look at what happens if we relax the standards. Instead of saying what in her evidence base lets her conclude that everything is present, let our presentist claim to know that presentism is the correct metaphysic of time based on holistic facts concerning overall theory choice.

This presentist claims to know that presentism is true, and *therefore* know that everything is present, since the latter claim just is one of the things that that theory says. And how is it that she knows that presentism is true? For the same reason we know any theory: it is justified as part of a holistic program of theory choice—it affords certain benefits, avoids certain costs, secures certain theoretical virtues, etc., etc. To respond thus would be to give up on trying to say something about how your particular evidence base rules out the scenario in which some things are non-present. This presentist is playing a different epistemic game: she is simply claiming knowledge of her theory of time—as a part of her overall theory of the world, which is justified on cost–benefit analysis, etc.—with her knowledge that she is present being obtained in virtue of being a part of that known theory.

This is to move to an externalist account of knowledge: the thought now is that presentism can be known in part because the extra-mental world cooperates. After all, we have no guarantee that the theory that is selected by pragmatic principles of theory choice is true; but the thought is that *if* the world cooperates and the theoretically best theory *is* true, then we can know it to be so. Thus facts that are inaccessible to us, concerning how the world external to us happens to be, play a role in our bases for knowledge.

I think this is indeed the way the presentist should respond. You do not try to say anything more about how you know you are present other than that this is what your best total theory says, and you do not say anything more in justification of that theory other than that it is the result of applying the algorithms of theory choice to it and its competitors. In particular, you reject any demand to specify some particular piece of evidence that rules out rival theories or that you could only be in possession of if this theory were true.

But if this is a good way for the presentist to respond to the skeptical puzzle, I think it is equally good for their A-Theorist rivals to respond this way as well. After all, *all* sensible A-Theories claim that this is the present time, not only presentism: so no matter what particular A-Theoretic metaphysic I settle on, my overall theory of the world will tell me that I am present. It is not as though the moving spotlighter, for example, simply thinks that the Now shines down on *some* time or other; it is part of her theory that it shines down on *this* time. That is not some ad hoc addition to their theory: A-Theorists are motivated precisely to capture the thought that there is something special about *this* time—the only reason they have for thinking that *some* time is privileged is that they think that *this* time is, so it is simply a part of the moving spotlight and growing block theories that we are present. To believe a particular A-Theory is to believe *de re of* some particular time that *it* is present, not simply to believe *de dicto* that *there is* some present time. So the moving spotlighter should, I think, simply claim that the reason she knows that she is present is precisely because her best theory of time says that she is so, just as our presentist has been moved to do.

Now, if part of what the moving spotlighter's or growing blocker's theory says is that it is *this* time that is present—as it must, if the above solution is to work—then, of course, she must constantly change her theory of the world, since *what* time is present changes. But that should come as no surprise. Every A-Theorist thinks that how reality is changes. So of course you have to change your theory of reality, to keep up with those changes. It is only the B-Theorist who can expect to settle on the best theory and then never have to give it up, for it is only they who think that how reality is is not something that is subject to change. (That is not to say that the B-Theorist does not believe in change: it is just that they think that change is variation across *portions* of reality, and deny that reality as a whole changes.) But for the moving spotlighter the theory will change in a predictable way at least: your theory of reality will tell you what theory of reality to have at any later time, since part of what your theory of reality does is tell you about what *will* be the case. So it is both unsurprising and unworrying that the moving spotlighter's best theory of reality is only ever momentarily true. (Things are tougher for the growing blocker, who needs to keep checking how the block has evolved and adopting a new theory of reality to account for that.)

One question remains if we are to have an adequate account of how the non-presentist A-Theorist can know that she is present. I have suggested that she knows that she is present because this is what her best theory of time says, and she knows this theory by having settled on it after correctly applying the principles of theory choice to it and its competitors. But *why* exactly is the theory that says that *this* time is present better than its rivals? I answer: precisely *because* it says that this time is present. That is a pre-theoretic datum: it is one of the beliefs we start from when we start to theorize about time. Now, of course, like any datum it can be revised if doing so leads to some greater theoretical benefit. There is no claim being made here that the presentness of *this* time is some kind of Moorean truth that a theory has to account for. But as with all common sense claims, it is a *pro tanto* virtue of a theory if it can account for it.

There is nothing question-begging about any of this. I am not claiming that our hypothetical moving spotlighter starts out *knowing* that she is present and relies on this knowledge to choose the best theory. I am not even relying on the claim that she is *correct* that it is *this* time that is present. I am only relying on the claim that it is a tenet of common sense that it is *this* time that is present and, hence, that it is a *pro tanto* virtue of a theory that it entails that this is so, and that this is a *pro tanto* reason to accept that theory against a rival theory that says that some other time is present.

Of course, choosing a theory because it is theoretically virtuous is fallible. There is no *guarantee* that the best theory is true, hence no guarantee that our moving spotlight theory that says that it is *this* time that is present is true. But that is acceptable: we were only after knowledge, not a guarantee. That we should be in this position—claiming knowledge, but also acknowledging the fallibility of this knowledge—is no surprise given the externalism about knowledge that we have ended up with. If you think such fallibility is a reason for skepticism then you probably have internalist sympathies: but in that case, as I argued in the previous section, you face a problem no matter what kind of A-Theorist you are. For my part, I am happy with externalism about knowledge: I think that if the world cooperates, then the theory that is selected as best will be true, and we will know it to be so, and hence we will know that what that theory says is true. And so, on the assumption that their theory is indeed true, presentists and non-presentist A-Theorists alike can know that they are present, since they can know that their best theories of reality claim that they are so.

Let me tackle an objection. I said above that the modal status of presentism would not solve the puzzle: that even if it is necessary that everything is present, this does not explain how we can know that we are present, for neither reliability of belief-forming process nor safety of belief is a sufficient condition for knowledge. But someone might object that one of them is a *necessary* condition for knowledge, and that this puts the presentist back in a better epistemic position.

So perhaps what I say is right about how to get in a *pro tanto* good epistemic position with respect to the claim that you are present—that is, that you should believe it because it is what is said by your best overall theory. Nonetheless, one might argue, the belief only amounts to *knowledge* if this epistemic goodness is not defeated. If presentism is true then that is so but, so the objection goes, if presentism is not true then considerations of reliability or safety defeat the epistemic goodness and hence the non-presentist's claims to knowledge are undermined.

But *is* the moving spotlighter's (e.g.) belief that she is present unsafe? Could it easily have been false? I do not think so. Exactly how we should formulate a safety constraint on knowledge is, of course, contentious, but I will go with the following:

> *Safety*: For all cases α and β, if β is close to α and in α A knows that p, then any counterpart of A's belief that p in β is true.

This takes Timothy Williamson's formulation[14] and modifies it in two ways suggested by David Manley.[15] Williamson demands that you not *falsely* believe in β; but as Manley points out, this does not seem sufficient, for there are other ways your belief could fail that, were they easy possibilities, intuitively render your actual belief unsafe. Suppose I form the belief that Robbie's first uttered sentence today will be true. Robbie is undecided between two options, but he is determined to do one of them: to start the day by saying "2+2=4" or to start it by saying "This sentence is false." He flips a coin and does the former. My belief, then, is correct. But intuitively, it is unsafe: had the coin landed the other way, my belief would not have been correct. But it would not have been *false*, it would have been paradoxical. So we have to demand that in the close possibilities I believe

[14] Williamson (2000, p128). [15] Manley (2007).

truly, not merely that I do not believe falsely.[16] The second modification taken from Manley is as follows.[17] Williamson demands that if you know p in α then if you believe *that very thing* in β, it can't be false in β. But intuitively your having a true belief can be lucky because you could easily have believed a *different* thing that is false. Suppose I am really bad at distinguishing dogs and foxes, and whenever I see one of either I form the belief that it is a dog. Outside my window are many foxes and one dog. I look at the dog and say "That's a dog." My belief is correct but, intuitively, it is not knowledge because I could easily have got it wrong. But I could not easily have got *that very belief* wrong: plausibly, it is a necessary truth, and could not have been wrong. The easy possibility is that I could have believed something *else* that is *close enough to my actual belief* and which would have been false, and that this renders my actual belief unsafe. Hence Manley's invocation of a counterparthood relation across beliefs: in the close situation in which I point to a fox and say "That's a dog" my belief is close enough to my actual belief to count as its counterpart, and that is why my actual belief does not constitute knowledge according to *Safety*.

I will not attempt to say anything particularly informative about when two beliefs are counterparts in the relevant sense. Indeed, I suspect that the best way of getting a grip on the notion is via thinking about what beliefs count as knowledge, just as Williamson thinks that the best way to get a grip on what cases count as close possibilities is to think about what we know. However, it is clear to see why one might naturally think that if the moving spotlight theory is true then my belief that I am present is not safe. The thought is that in the *actual* situation there are false counterparts of my belief that I am present, such as Caesar's belief that *he* is present. Caesar is in a relevantly similar epistemic situation to my own, and his belief has the same character as my own, so it looks like a good candidate for being a counterpart to my belief that I am present; and so, since on any account actuality must be close to itself, this means there is a close possibility where I have a false counterpart belief, and hence my belief is unsafe. Or alternatively, hold fixed that the counterpart belief is my believing that I am present and move the spotlight forward or backward to a time after my death or before my birth: if that case is close to actuality then my belief counterpart is false in a close possibility and hence unsafe.

[16] Manley's demand is a bit different: that in close possibilities my belief not *fail*, where one way to fail is to believe falsely, another is to believe something paradoxical, another is to believe something with gappy content, etc. Manley's option only diverges from mine here because he wants to allow for certain theoretical options that I am happy to ignore and which are not relevant to the current issues. See Manley (2007, p405).

[17] Cf. Hiller and Neta (2007).

But I think this is a mistake. Consider the following scenario. It is September 17, 2014, the eve of the referendum on whether Scotland should be independent of the United Kingdom. Hamish considers the following two claims.[18]

(i) In all close possible circumstances in which Caesar crosses the Rubicon, the immediate future contains a Roman civil war.
(ii) In some close possible circumstance in which Caesar crosses the Rubicon, the immediate future contains a referendum on Scottish independence.

Intuitively, Hamish's thought that (i) is true and his thought that (ii) is false. There is no close possible circumstance in which Caesar crosses the Rubicon and the immediate future contains a referendum on Scottish independence. And yet, Hamish's *actual circumstances* are such that Caesar crosses the Rubicon and that the immediate future contains a referendum on Scottish independence—and surely the actual circumstances are at least amongst the closest possible circumstances.

I think the lesson to take is that the moving spotlighter should think of a possible circumstance not as a world but as a world–time pair, and when we are to think about close possible circumstances in which something happens, we are (often[19]) to think of collections of world–time pairs in which that something *presently* happens: that is, we consider circumstances in which the spotlight falls on the time at which the event we are considering is happening. When thinking about what happens in close possible circumstances in which Caesar crosses the Rubicon, we should think about world–time pairs in which the spotlight shines on Caesar's crossing the Rubicon. This does not include the actual circumstances, where Caesar crosses the Rubicon two millennia before the time on which the spotlight shines. That is why the immediate future in such circumstances contains a Roman civil war but not a referendum on Scottish independence: since the spotlight has, in those circumstances, shifted back to the time at which Caesar made the crossing, the immediate future of those circumstances is what happens immediately after *that* time, not what immediately happens after the time on which the spotlight actually shines.

[18] Historical background: many historians think that Caesar's crossing the Rubicon was a pivotal event that led to the subsequent Roman civil war. Assume this is correct for the sake of the example.
[19] I say "often." I suspect that one can create weird conversational contexts in which this is not true. One can easily hear this as true, I think: "In close possible circumstances in which I meet Robert the Bruce, I have traveled in time and am in what would be the objective past." But here we are using the talk of time travel to force a reading whereby we keep the spotlight on the actual present, and move ourselves back into the actual past to when Robert the Bruce is. My claim is simply that this is an abnormal reading: the more natural reading—and the one relevant to assessing the safety condition—is where we move the spotlight to the time at which the event happens.

So now consider the close possible circumstances in which I believe that I am present. In all of them, I *am* present, precisely because I consider only possible circumstances in which the spotlight shines on the time at which the event I am considering—my believing that it is present—happens. My belief that I am present *could not* easily have been false, then: the close possible circumstances in which I believe it are ones in which the time at which I have the belief is the time on which the spotlight falls, which is exactly what *makes* my belief true in those circumstances. Similarly, the counterfactual "Were I to believe that I was present, I would be present" is, I suggest, true in ordinary contexts, just because evaluating the counterfactual ordinarily requires considering as closest exactly those scenarios in which the spotlight shines on the time at which I have the belief.

Likewise if Caesar's belief that he is present counts as a counterpart of my belief. The threat was that here we have *in actuality*—as close a possibility as you can get!—a false counterpart to my true belief. But in fact, this is a mistake. What matters is not whether there is a false counterpart belief, but rather whether there is a close possibility in which the counterpart belief is had and is false. And the close possibility in which Caesar's belief is had is *not* the world–time pair of the actual world and the actual present, but rather the world–time pair of the actual world and the time at which Caesar has the belief. The counterpart belief is actual, and it is false; but the close possible scenario in which the belief is had is one in which the counterpart belief is true, precisely because in the close scenarios the spotlight is moved to the time at which the belief is being believed, and the spotlight being there is what makes the belief true. My belief that I am present could not easily have been false, because were a belief with that character to be had, it would be had in what would be the present, and hence would be true.

And similar remarks apply to reliability. Consider the process: coming to believe that you are present as a result of learning that this is a consequence of the theory of time that maximizes theoretical virtues. If presentism is necessarily true then it is uncontroversial that this process is reliable. But, so the objection goes, if the moving spotlight theory is true then this process is unreliable, for there may be many merely past people who did just that, who falsely believe that they are present. But *is* this process really unreliable if the moving spotlight theory is true? If the moving spotlight theory is true, it is always the case that *when* someone goes through this process, the belief formed as a result of that process is true. Merely past people falsely believe that they are present; but their belief was true *when* they formed it—so at the time they implemented the above process, it did not make them wrong. And for the reasons given above, this will be modally stable: *were* someone to implement this process, it would lead them to a true belief. What more do you want from a reliable process other than that whenever someone implements it, it goes right, and that were someone to implement it, it would go right? That is reliability par excellence! We are tempted to think that the process is unreliable, given the moving spotlight, because there *are* (atemporal "are") lots of instances of the process where the resulting belief is (atemporal "is") false. But I think that is mistaken. There is nothing wrong with the *process*: *when* it was implemented, it got it perfectly right—the belief *was* true at that time. The reason that those instances of the processes are ones where the belief is false is not a result of the unreliability of the process but rather is a result of the instability of the truth of the belief. The process is perfectly reliable: implement it, and you are guaranteed to get a true belief when you do! But because what is true *changes*—and beliefs about what is present do so more rapidly than anything—you cannot count on the belief that you form as a result of undergoing this process *staying* true. But that is no problem: we *do not* expect it to stay true that we are present. We merely want to know that it is so when we have the belief: and since the process I recommend is guaranteed to lead to a belief that is correct at the time you form it, I think there is no problem here.

Another way you might argue that the knowledge of one's presentness is defeated if the moving spotlight is true is by appealing to this principle:[20]

> *Statistical Knowledge*: If you believe p on basis E then, if most of the people who believe p on basis E believe falsely, you do not know p on basis E.

Since the moving spotlighter we have in mind, but not the presentist, thinks that most people who believe that they are present believe falsely,[21] this would give us an argument that if the moving spotlight theory is true then we do not know that we are present—an argument that would not trouble the presentist who does not think anyone believes falsely that they are present.

But I do not think we should accept *Statistical Knowledge*. I think it only sounds appealing because *Safety* is appealing; and it is easy to think that if most people who believe that p on basis E do so falsely then your belief that p on basis

[20] This was suggested to me by Jonathan Tallant.

[21] Actually, it is not quite this quick. For while the moving spotlighter thinks that many, many people believe falsely that they are present, it does not follow that many people believe falsely on the relevant basis. While I would like to think that many future people are going to be convinced by the arguments in this chapter and believe that they are present on the basis that I am recommending, it may turn out that I am the only person in history who believes that they are present on the recommended basis. But even were the antecedent of the second conditional in *Statistical Knowledge* false for this reason, that would be of little consolation, for surely it is not the case that I know that I am present simply because of an accident concerning how popular my proposed basis for believing that you are present has been at other times.

E must be unsafe. But as I have argued, our belief that we are present *is* safe, no matter whether *Statistical Knowledge* is satisfied.

The view I am defending is heavily externalist. I am relying on the world doing its part for the moving spotlighter to come to have knowledge that she is present. The moving spotlighter makes all the right epistemic moves: she believes a theory on the basis of applying good principles of theory choice, and she goes on to believe what she knows to be a consequence of that theory. Internally, she is doing everything right: but unless the world plays its part—unless the good principles of theory choice indeed select the true theory—her belief will not amount to knowledge. And on reflection, *Statistical Knowledge* is obviously not a principle the externalist should accept. Forming a belief that p on basis E might be a good epistemic move. So *everyone* that believes that p on E is in an internally good epistemic position: they are all *primed* for knowledge, in that they have done everything they can and only need the world to play its part. But if p is a claim about one's *position* in the world, then it is clear that the world might play its part for some of the people who believe p on E but not others: indeed, that it might play its part for only a select few, granting *them* knowledge that p even though most people who believe p on E lack knowledge. And so it is easy to see why the externalist should not accept *Statistical Knowledge*. In our present case, p is the claim that you are present: it is precisely a claim about one's position in the world—people who believe it are believing themselves to be at the unique objectively privileged time. If I am right then *everyone* who believes that they are present on the basis I offer them is doing something epistemically praiseworthy: they are each warranted in their belief. They are all primed to know that they are present, if only the world cooperates. But it only cooperates for relatively few of them: so while those few know that they are present on that basis, most who believe that they are present on that basis will not know that they are present (since they are not). This means, of course, that there is an epistemic difference between people without there being any relevant difference in their accessible evidence, since how things appear to be is the same in all relevant respects to all these people who are primed for knowledge of their own presentness. But that should be of no concern to an externalist: only an internalist will think that all the epistemic features of a situation supervene on the facts about what evidence is accessible to whom. Externalism is precisely the view that how the external world is can make a difference: in this case, the feature of the external world that makes the difference to the epistemic facts is which time is present.

If what I have said is correct, I have explained how the A-Theorist—presentist and non-presentist alike—can have knowledge that they are present. Of course, I relied on not uncontroversial views about knowledge in making my case: my account was heavily externalist, and an internalist about knowledge will not be convinced. I am unworried: what was troubling was the thought that non-presentist A-Theories ruled out knowledge of one's presentness simply by virtue of their metaphysics. The objection made by Bourne, Braddon-Mitchell, et al. was not presented as relying on any substantive theory of knowledge: the objection was that simply the combination of many times and an objective present was enough on its own to rule out knowledge of one's presentness. I am happy if I have shown that to be false: at least on some theories of knowledge, the people who are present can come to know that they are present, even if the moving spotlight or growing block metaphysics are true. I happen to find the required views about knowledge plausible, but that is another story. And if this externalism about knowledge is too lenient then it is not clear how *any* A-Theory allows for our knowledge that we are present. Such knowledge looks unobtainable, given such high standards, even if presentism is true. So on pain of skepticism, all A-Theorists should embrace externalism, in which case no A-Theorist has a problem.

But now a confession: what I have *not* done is show that everyone who is present has this knowledge. The route to knowing that you are present required settling on a theory that says that you are present: moreover, it required settling on that theory for sophisticated reasons—that it is the overall best theory of reality, maximizing the balance of costs versus benefits, etc. The thought then is that if the world cooperates—if it is such that the most theoretically virtuous theory is *true*—then those of us who applied the virtues *correctly* and settled on the *true* theory will thereby *know* that it is true, and hence be in a position to know that they are present (since the theory says exactly this). But what of the poor people who have chosen no metaphysic of time? How do they know that they are present?

I think someone who accepts the above defense has simply to bite the bullet, and say that if the A-Theory is true then such people—the vast majority—simply do not know that they are present. Whether I have solved the epistemic puzzle, then, depends on what you think the datum is. If it is that present people all (or mostly) know that they are present, then I have not accounted for that. But I think the only datum is that we are in a *position* to know that we are present. What seems objectionable is if knowledge that we are present is *in principle* beyond our reach: that is what Bourne, Braddon-Mitchell, et al. charge non-presentist A-Theories with. I have argued that such knowledge is within our reach, whatever the correct metaphysic of time; and that, I think, is enough. It is no problem if such knowledge is not widespread, so long as we know how to obtain it. If one goes this route, then, a certain type of metaphysical evangelism is called for. Just as some want to convince you to believe a certain religious theory, promising you sure and certain knowledge of your salvation if you only go ahead and believe, we temporal metaphysicians should do our best to convince the folk to sign up to our preferred metaphysic of time, promising potential converts that knowledge of their presentness awaits if only they believe for the right reasons.

1.4 In Praise of Something Higher than Knowledge, and Against Metaphysical Idlers

The views which *prima facie* have the hardest time with respect to the epistemic puzzle are ones which posit non-present entities and which take them to be having the experiences and beliefs that they had in the past. I have argued that even given such a view we can solve the epistemic puzzle, at least if we are willing to accept a liberal externalism about the standards for knowledge. And if stricter standards are demanded, then there is just as much a problem for the presentist as for the non-presentist A-Theorist.

But while such views can solve the epistemic puzzle, there nonetheless seems something problematic in the claim that all these past and future people falsely believe that they are present. What is missing is a sense of the metaphysics *guaranteeing* our being present. For the presentist, the very nature of time guarantees that one is present, because that is the only way to be. Now, this is not an *epistemic* guarantee: the truth of presentism does not render the claim that you are present certain or beyond doubt or anything like that because, as has been belabored in the preceding discussion, the metaphysical theory that secures your presentness is itself not certain or beyond doubt. But nonetheless, the presentist can at least say that *given* the nature of time, there is simply no option but that you are present. The moving spotlighter or growing blocker, if she follows the path offered her in section 1.3, cannot say that. Her total theory of what time is like entails that she is present—since, as we have just seen, it is simply part of the particular A-Theory that she accepts that the spotlight falls on this very time—but the *nature* of time is, on her view, perfectly compatible with her being non-present. She recognizes the coherence of a theory of time that agrees with hers exactly on what reality is like except that it places the present elsewhere in history. She simply thinks that this theory is less good and ought not to be accepted; it does less well when it comes to weighing up the costs and benefits that are relevant to theory choice.

This may well, as argued in the previous section, let her *know* that she is present. But nonetheless, something feels wrong. Intuitively, that these goings on—the ones we currently have direct experience of—are present should be guaranteed by the very nature of time. You might not know you are present because you do not know what time is like but, intuitively, *once* you know what time is like there should be no more room for doubt as to whether or not you are present. Your knowledge that you are present should be secured by your knowledge of the very nature of time, whereas the account offered in the previous section makes your presentness wholly unconnected to the *nature* of time: that you are present is simply an extra fact about the world, one that you ought to believe simply because it is advantageous if true.

I think there is both an epistemic and a metaphysical advantage afforded to views that have your presentness guaranteed in some sense by the very nature of time. To see the epistemic advantage requires one to be somewhat Aristotelian in one's epistemology. In the previous section I argued that someone who believes in Caesar, and who thinks that Caesar is thinking that he is present, can nonetheless know that they are present. But for Aristotle, there is more to epistemic goodness than mere *knowing*. I can go out and see a white swan and form the belief that there is a white swan on that basis, and (provided, at least, that nothing funny is going on with my perceptions, etc.) as a result I can *know* that there is a white swan. But this is not the ideal epistemic state to be in with respect to that proposition: I can know it *better* by deducing it from general principles about the essences of things.[22] For Aristotle—and I am very sympathetic to this claim—there is not one sole epistemic virtue, knowledge, that you either have or lack with respect to a proposition, and that's the end of the epistemic story: there are better and worse ways to know things. The ideal epistemic state might not be attainable by us with respect to any proposition, and it is certainly not attainable by us with respect to all propositions, but it is an ideal to which to aspire.

If Aristotle is right that to know something on the basis of a deduction from a general principle about the nature of things is a better way of knowing than, say, to know it on the basis of an inductive or abductive argument, then we can see why there is, after all, an epistemic shortcoming to the metaphysics of time that entail that Caesar exists and believes himself to be present. It is not that such views entail the absurd consequence that we do not know that we are present; it is that they entail the disappointing consequence that we can only know it as the result of the abductive argument given in section 1.3. This is disappointing because it means that we cannot know that we are present in the best way: it leaves the epistemic ideal to which we ought to aspire unobtainable. That is no reductio of such a metaphysic; as Aristotle himself allows, we fail to meet such an

[22] See Pasnau (2014) for discussion.

ideal all the time, and this is not particularly worrying: merely knowing that p is enough to get us by in our ordinary lives. But insofar as there are rival metaphysics of time that allow us to obtain *better* knowledge, that is a *pro tanto* mark in their favor. And there are such rival metaphysics: presentism—and indeed, any metaphysic that holds that the nature of time guarantees that all who think that they are present *are* present—allows for this more ideal epistemic state, for then we can deductively infer our presentness from claims about the very nature of time.

There is also a metaphysical disadvantage to views that have Caesar believing that he is present and being in a similar evidential position to ourselves. If Caesar and I are both having certain experiences, completely alike in relevant character... well, while I can know that mine are present and that Caesar's are not, as previously argued, the problem is that the presentness of the experiences is not *doing anything*. The property of being present is a metaphysical idler: having it or not is not making any difference to the character of the events in question, or the experiences of them. *Who cares* if I know that my current experiences in fact have that property, then, for the property is rendered entirely uninteresting! Presentness is, on such a view, a mere epiphenomenon, of little metaphysical interest. We have a moving spotlight that does not actually make things brighter.

And so I think the non-presentist A-Theorist had better say that there *is* a difference between Caesar's situation and our own: either that Caesar lacks experiences, or that his experiences are different in kind to our own, or something like that. Not so that we can point to this difference to show how we can *know* that we are present as opposed to Caesar, but rather so that we can point to this difference to show *what work* the property of being present is doing in the world: what *difference* it makes to the things that have it. If we cannot point to some such work—if we cannot say what difference it makes to an event that it is present—then an event's having that property is a totally uninteresting feature of it. There may be no problem in coming to know what events have that feature, but that is of little interest if it is a boring thing to know. There is no problem with how we can know whether we are present, but if this is all presentness amounts to, there is no reason to *care* either.

In the case of the moving spotlight, which is my concern in this book, this metaphysical problem—that the property of being present is a metaphysical idler—arises due to a particular conception of the moving spotlight metaphysic. We naturally think of the moving spotlight view as taking a B-Theoretic metaphysic and *adding* something to it, namely, that one moment in history have this special property of being present. On this way of thinking about things, the moving spotlighter believes in all the same things the B-Theorist believes in, and for every way the B-Theorist believes something to be the moving spotlighter also believes it to be that way, but she adds to the account an additional feature: that some of those things have the property of being objectively present. In that case, all that changes in the world as time progresses is which things have the property of being present: the other features of things remain constant from one time to another, just like they do for the B-Theorist. In that case, the property of being present really is a metaphysical idler: it makes no important difference to the things that have it. The question with that is not how we can know what has it, the question is: why would we have any reason to posit such a feature of reality? It would be like believing in absolute rest in a Newtonian world.

The goal in the chapters that follow will be to develop a version of the moving spotlight that rejects this way of thinking about things. On the metaphysic I will defend, we do not simply take the B-Theorist's metaphysic and *add* something to it. The moving spotlight view I advocate views the world in a fundamentally different way than the B-Theory does: the *whole* of reality changes from moment to moment. Caesar, myself, and the first lunar colony all exist, and always did and always will exist, but we are all in constant flux: we are each different from how we were yesterday, and we will each be different again tomorrow. All things—past, present, and future—are constantly changing.

In particular, as we will see in chapter four, on the moving spotlight view I defend, the only way things are, simpliciter, is the way they are *now*. Caesar is now some way, on my view, but he is not now experiencing anything, nor is he now having any beliefs (*a fortiori* he is not believing that he is present). As a result, he lacks experiences and beliefs *simpliciter*. Likewise for every non-present entity. So the only people who believe that they are present, on my view, *are* present. Now, this is *not* what solves the epistemic puzzle; as above, all this means is that my knowledge that I am present is on as good or as bad a footing as my knowledge of my metaphysic of time. What solves the epistemic puzzle is the externalist story given above: I know I am present because my metaphysic tells me so, and I know *it* because it is the best overall theory in terms of maximizing benefits and minimizing costs (as the arguments of the subsequent chapters aim to show). Believing that Caesar et al. lack beliefs does not solve the epistemic puzzle, but what it does do is allow for the epistemic *benefit* that we can know that we are present as a result of deducing it from general principles about the very nature of time. And it solves the *metaphysical* problem: the property of being present is clearly no metaphysical idler on this view—it makes a significant difference to how its bearers are. Caesar is not experiencing anything, but he *used* to have certain experiences, and the difference is that he used to be present: presentness makes a difference, and it is clear why it is a difference we ought to care about.

This is, of course, similar in some respects to Forrest's view. However, my metaphysic is more extreme than Forrest's, and as a result, I think, less ad hoc. Forrest also denies that Caesar is having any conscious experiences. But every *other* way that Caesar was, he remains. Being present makes a difference, but *only* to whether something is conscious. In all other respects, Caesar remains as he was, he simply ceases to be conscious once he is non-present. Caesar has all the properties that one might have thought that consciousness *supervenes* upon, but he is not conscious. That strikes me as ad hoc: it seems that the only reason Forrest has for denying that Caesar is conscious is because he wants to avoid the epistemic objection. (Of course, I do not think he succeeds in doing that; but even if this were to solve the problem, it seems ad hoc.) By contrast, as will become clear in chapter four, I think that Caesar is now *completely* different from how he used to be: not only is he no longer conscious, but he no longer has any of the ordinary properties that we would think are relevant to something's being conscious. Caesar's being conscious, along with his being a certain height, mass, and having a certain arrangement of neural firings in his brain, etc., are merely ways things were, and they are no part of how reality is.

The challenge with such a view is how past ontology is meant to make true historical truths if how things were in the past is no longer a way reality is. *Prima facie*, part of the point of believing in the past is to serve as an ontological record of how things were: that the way things were remains a part of how reality is, to serve as truthmakers for claims about what used to be the case. But as we will see in the next chapter, this principle simply has to be abandoned, for if it is accepted in full generality it results in a vicious version of McTaggart's paradox. Even if past entities still exist, there must be at least *some* respects in which they are no longer as they were. And once we allow that past entities are different in some respects from how they were when they were present, there is no in principle objection to an account like mine that says that they are very different from how they were when they were present. And as we will see in chapter four, I think we can still have a satisfying account of how past entities make true historical truths, despite now being different from how they were.

2

From A-Theory to Presentism?

Part 2: McTaggart's Paradox

McTaggart argued that the A-Theory leads to inconsistency. I argue in §2.1 that he was wrong, but that there is a serious McTaggartian challenge to non-presentist A-Theories: an attractive view on how tensed truths depend on reality leads to inconsistency. In §§2.2–2.3 we look at the analogy between time and modality. In §2.2 we will see how the presentist can respond to a version of McTaggart's argument put forward by Nicholas J. J. Smith by making a distinction between a time t being a certain way and t's being that way *at itself*. This is a familiar move from actualist theories of possible worlds, where there is a distinction between how an ersatz world is and what's true *at* it (i.e. the way it represents things as being). In §2.3 we see Phillip Bricker making a similar move to respond to the modal analog of McTaggart's argument as applied to his view that there are many genuine possible worlds, one of which is objectively special. I argue that there is a disanalogy between Bricker's theory of modality and the moving spotlighter's theory of time that prevents the latter from making the analogous move in response to McTaggart: while Bricker's view is that reality is an ultimately non-modal place, the moving spotlighter's view is that reality is an ultimately temporal place. Following Bricker's solution, I argue, would yield the view that really there is a *stuck* spotlight, but we can speak *as if* the spotlight is moving. We end §2.3 with a challenge to the moving spotlighter: she must solve McTaggart's paradox, while still giving a satisfying account of how tensed truths depend on reality, and she must be able to distinguish her view from the stuck spotlight view with a fancy semantics. This challenge is further borne out by considering Kit Fine's version of McTaggart's argument in §2.4. The challenge yields constraints that must be met by the metaphysic to be defended in chapter four.

2.1 McTaggart's Argument for the Inconsistency of the A-Theory

In the last chapter I argued that there is no reason from considerations of epistemology for the A-Theorist to adopt a presentist ontology. In this chapter we will investigate whether combining the A-Theory with a non-presentist ontology leads to metaphysical problems. In particular, we will be looking at McTaggart's argument that the postulation of A-properties generates an inconsistency. Now of course, McTaggart's argument was directed against *all* A-Theories, not merely non-presentist A-Theories.[1] And so our first item of business will be to dismiss this objection; we will proceed later to look at a version of the objection that targets only non-presentist A-Theories.

A thorough discussion of McTaggart's argument that took into account the massive literature that it has spawned, looking at every different attempt to present a version of the argument and getting to the bottom of each, is a task that is far beyond what can be accomplished in a chapter. That would take a book in itself. And so we must make a decision about how to limit our attention, and inevitably we will leave good work undiscussed. I will choose to concentrate on McTaggart's original text, for obvious reasons, and on a recent attempt at presenting a McTaggart-esque argument by Nick J. J. Smith, which I think is interestingly different from McTaggart's argument and which, being recent, is less thoroughly discussed in the literature. I will ignore famous McTaggart-esque arguments presented by D. H. Mellor[2] and by James Van Cleve,[3] both because much ink has already been spilled on them, and I do not have much to add to that, and also because those arguments are pretty similar to McTaggart's and I do not myself think that there is something to be learned from seeing why they fail that cannot be learned just from discussing McTaggart. I will end by looking at Kit Fine's recent McTaggart-inspired argument, which he uses to motivate a non-standard realism about tense.

Let us begin, then, with McTaggart:

> Past, present, and future are incompatible determinations. Every event must be one or the other, but no event can be more than one. If I say that any event is past, that implies that it is neither present nor future, and so with the others.... But every event has them all [if it has any]. If M is past, it has been present and future. If it is future, it will be present and past. If it is present, it has been future and will be past. Thus all the three characteristics belong to each event. How is this consistent with their being incompatible?[4]

McTaggart is arguing that if the A-Theorist is correct in thinking that anything is objectively past, present or future then everything that has one of those three properties must have all three. But nothing can have all three: any two of them are incompatible. Hence, the A-Theory is impossible.

[1] In fact, it was directed at *all* theories that said there was time. But that is only because McTaggart thought that in order for there to be time, there had to be an A-series of events in reality. I will ignore this part of his argument: both because since I am assuming the A-Theory, it is irrelevant, and because it is pretty much universally thought to be wrong.

[2] Mellor (1998). [3] Van Cleve (1996). [4] McTaggart (1908, p468).

One quick comment before we proceed. It is a mistake[5] to say that every pair of the three A-properties is logically incompatible: something *can* be each of past, present, and future, if time is circular. If I could walk forward round a circumference of the Earth, I would arrive back where I started, and likewise if I walked backward round that circumference, and so my starting point is both here, ahead of me, and behind me. That is possible because the line I walk is the circumference of a circle. Similarly, if time is circular, then once we go far enough into the future we will arrive at what is now the past, and from then continue on back to our starting point: and so our present is also our future and our past, and so this time has all three A-properties. I see no reason to deny the possibility of circular time, hence no reason to claim that the A-properties are logically incompatible. Nevertheless, while something *can* have all three A-properties, it is evident that they *need* not all be had together: it is *possible* that something is present without being past or future, etc. Since McTaggart's argument aims to threaten even the possibility of only one A-property being had, it needs to be taken seriously. For ease of presentation, however, I will continue to speak of the A-properties being incompatible: understand this, if you like, as simply meaning that they are incompatible in the actual world, given that time is not actually circular.

Now, the first reaction one has to McTaggart's argument is that he has made an obvious mistake. The troublesome conclusion is that M *is* both present and past (or both present and future, or etc.). But all that follows is, as McTaggart himself puts it, that M is present and *was* future and *will be* past, or that M is future and *will be* present and *will be* past, etc. The incompatibility of the properties means that nothing can have them *at the same time*; but there is no problem, one would naturally think, in something's having incompatible properties *successively*.

McTaggart, as is well known, anticipates this objection. He offers two responses: that it is circular, and that it leads to vicious regress. Here is the charge of circularity, which has generated less subsequent discussion.

> This explanation involves a vicious circle. For it assumes the existence of time in order to account for the way in which moments are past, present and future. Time then must be pre-supposed to account for the A series. But we have already seen that the A series has to be assumed in order to account for time. Accordingly the A series has to be pre-supposed in order to account for the A series. And this is clearly a vicious circle.
>
> What we have done is this—to meet the difficulty that my writing of this article has the characteristics of past, present and future, we say that it is present, has been future, and will be past. But "has been" is only distinguished from "is" by being in existence in the

[5] One that Smith also makes. Smith (2011, §2.1).

past and not in the present, and "will be" is only distinguished from both by being in existence in the future. Thus our statement comes to this—that the event in question is present in the present, future in the past, past in the future. And it is clear that there is a vicious circle if we endeavour to assign the characteristics of present, future and past by the criterion of the characteristics of present, past and future.[6]

Here, McTaggart is complaining that we have appealed to resources whose legitimacy is in question in order to defend their legitimacy. Whether or not it is coherent to attribute A-properties to things is exactly what is up for debate. McTaggart has presented us with an argument that the postulation of such A-properties leads to incoherence: to us attributing to things properties which are incompatible. We respond that the incompatible properties are had *successively* not simultaneously, and hence there is no problem; but, McTaggart objects, to claim that they are had successively is just to claim that while one is had *now* the other is had merely in the past (or future)—and that is to simply make another attribution of A-properties, whose legitimacy is what is up for debate. Hence, our response to McTaggart presupposes the legitimacy of A-properties in order to establish their legitimacy; and this, McTaggart says, is viciously circular.

What ought we to make of this charge? McTaggart is surely right that we are appealing to the legitimacy of what is being attacked when we come to their defense. The question is: is that dialectically inappropriate? I think not. Let us look at a case that I think is analogous to McTaggart's.

Kripke, channeling Wittgenstein, argues that there are no meaningful facts because there is nothing to determine which of infinitely many rival meaning hypotheses are correct. Nothing about the world determines that I mean addition rather than quaddition by "addition," for example.[7] In arguing for this Kripke tries to show that various ways of choosing between the rival hypotheses will not work. One way he considers is that it is the simplest hypothesis that is the correct one. In response, he says:

[A]n appeal [to simplicity] must be based either on a misunderstanding of the sceptical problem, or of the role of simplicity considerations, or both. . . . [S]implicity considerations can help us decide between competing hypotheses, but they obviously cannot tell us what the competing hypotheses are. If we do not understand what two hypotheses *state*, what does it mean to say that one is "more probable" because it is "simpler"? If the two

[6] McTaggart (1908, pp 468–9).

[7] Where quaddition is a function that yields the same values as addition for numbers small enough to have been considered in the course of human history, but which yields aberrant values given as input numbers greater than any that will ever be considered.

competing hypotheses are not genuine hypotheses, not assertions of genuine matters of fact, no "simplicity" considerations will make them so.[8]

Kripke's point, I take it, is this. Grant for the sake of argument that if we have multiple hypotheses about what is meant by a term then the simplest of those hypotheses is the best one. Still this is of no help, because the conclusion proper of the skeptical argument is that there *are* no such competing hypotheses because any hypothesis about the meaning of a term is literally contentless. As Alexander Miller puts it, "if two ascriptions of meaning do not have truth-conditions, what does it mean to say that one of them is more probably true because it is simpler?"[9]

I think that Kripke's objection here is confused. Consider the dialectic that was intended to establish the skeptical conclusion. We were given a challenge by the skeptic to account for why it was we meant addition by "plus." And, the argument went, *if* we cannot answer the challenge, then we are stuck with a non-factivity about meaning. The simplicity considerations are raised as an attempt to *answer* that challenge, so we cannot argue against that attempt by appealing to a result which is only established if that attempt fails. Kripke is illegitimately assuming the conclusion of his skeptical argument at a point in the dialectic where it is still to be established. If simplicity considerations are such that they can adequately choose between rival hypotheses then the skeptical argument fails and the non-factivity of meaning is never established. One cannot argue against this attempted solution by arguing that the rival hypotheses lack content; that is to assume what, at this stage in the dialectic, is still very much up for debate: namely, the truth of the skeptical conclusion. So Kripke's argument against the appeal to simplicity here is unconvincing: he is assuming his conclusion at a point in the dialectic when it has not been established—at a crucial point, furthermore: a point at which it will *not* be established if the objection is successful. As Crispin Wright puts it:

Whatever criterion of preferability among competing hypotheses we come up with, its application can be appropriate only if we do genuinely have competing *hypotheses*, only if there is some "fact of the matter" about which we are trying to arrive at a rational view. Therefore—or so Kripke's thought presumably runs—we beg the question against the sceptic in appealing to any such criteria at this stage. But this surely gets everything back to front. It is only *after* the sceptical argument has come to its conclusion that the sceptic is entitled to the supposition that there is indeed no such fact of the matter. In the course of the argument, *he* cannot assume as much without begging the question.[10]

[8] Kripke (1982, p38). [9] Miller (1998, p173). [10] Wright (2001, p109, fn.6).

So I think Kripke's response here to the simplicity solution is mistaken. If you are trying to argue that p, and the success of your argument relies on your opponent being wrong about q, you cannot rely on the truth of p when arguing that q is false. That is to make use of your conclusion before you have a right to it.

This is not just a matter of each theorist making arguments that are good by their own lights but not by their opponents' lights. The charge here against Kripke is not that he is making an argument that his opponent, by their own lights, ought not to accept. The charge is that Kripke, by *his own lights*, ought not to accept the argument he is making.

Compare another debate. Lewis says that to be possible is to be true at some isolated concrete spacetime. Some object as follows: what does being the case at some concrete spacetime have to do with what *could* have been the case? What do concrete spacetimes have to do with *possibility*?[11] Lewis replies: the modal operators are quantifiers over concrete spacetimes, *that is* what they have to do with possibility.[12] Now clearly, that is not going to satisfy his opponents, since they will not agree that the modal operators *are* quantifiers over concrete spacetimes. So what to make of this debate? There is clearly *some* sense in which Lewis's response to the modal irrelevance objection is circular: he is trying to say what concrete spacetimes have to do with modality, and says something that nobody would accept unless they already thought that possibility concerned happenings at concrete spacetimes—for what possible reason could there be for thinking that the modal operators are quantifiers over concrete spacetimes other than that these are the things such that the truth of p at one of them amounts to p's being possible? So Lewis is relying on the truth of his account of modality when he responds to this objection to it. As Charles Chihara says:

> [Lewis's] response to the Modal Irrelevance Objection was that he had already explained what worlds have to do with modality by saying that the modal operators are quantifiers over them. But that is just what is being disputed.... So, from the point of view of his disputers, Lewis was simply begging the question against his opponents.... Lewis responds to [the] objection from the perspective of one for whom the analysis is already beyond question.[13]

However, one could charge Lewis's opponents with the same circularity: they object that concrete spacetimes have nothing to do with what is possible, but that is only the case if Lewis is wrong about what possibility is like, which is precisely what they are arguing for. So in some sense of "circularity," both participants in this debate are engaged in a circularity. But the question is: is either party doing

[11] See Chihara (1998), Jubien (1988), van Inwagen (1985).
[12] Lewis (1986, p98).
[13] Chihara (1998, p95).

anything dialectically illegitimate? I think not: both parties are saying things that are perfectly appropriate given their own background beliefs. Lewis's opponents here have prior beliefs about what modality is like that they are refusing to give up: what *could* be the case, on their view, cannot simply be a reflection of what *is* the case, no matter how extensive reality turns out to be. By contrast, Lewis is guided by the desire for a reductive Humean[14] theory of modality: what could and must have been the case supervenes on what matters of fact obtain, it does not *constrain* such facts.

What we have here is just a clash of views. Each party is putting forward arguments that are perfectly appropriate by their own lights; but neither should expect their opponent to be convinced by these arguments, since they should be able to recognize that their arguments are not good by their opponent's lights. There is nothing fallacious about this: in making *any* argument we make certain philosophical presuppositions (even if it is only a matter of what the correct logic is), and while we can only expect to *convince* people who share those presuppositions, there is nothing dialectically inappropriate in our putting forward such an argument against views that deny those presuppositions. It is fine to object to dialetheism on the grounds that there cannot be true contradictions, although we should not expect this to convince the dialetheist.

What is going on in the case of Kripke's Wittgenstein, however, is different. It is not that Kripke has a case for skepticism about meaning that is good by his own lights but not his opponents'. Kripke's case for skepticism, *by his own lights*, relies on there being nothing in reality to select one hypothesis about what a word means over rival hypotheses as being the *correct* hypothesis. By his own lights, if simplicity considerations can so select between rival hypotheses then his case for meaning skepticism fails. By his own lights, then, the claim that the rival meaning hypotheses are contentless is only established if it is first established that simplicity considerations do not select the correct hypothesis from the many candidates—and so Kripke himself should recognize that it is dialectically inappropriate to rely on the claim that the rival hypotheses are contentless when arguing that simplicity considerations cannot select between them.[15]

The kind of circularity Kripke is engaged in is fallacious, whereas the kind of circularity Lewis is engaged in is not. Lewis is merely doing what we of necessity

[14] "Humean" in the same sense as his account of laws is Humean: i.e. not one that was actually held by Hume, but one in which the facts about laws/possibility are merely a reflection of the underlying facts concerning what local matters of fact obtain.

[15] For the record, Kripke offers other arguments against simplicity considerations being appropriate for this job. But that is irrelevant for our current purposes: all that matters for the moment is the fallacy Kripke is engaged in when he dismisses simplicity considerations for *these* reasons.

do: making certain presuppositions that the success of our argument depends on. The stronger those presuppositions, the more limited one's audience of potential converts is, but there is no fallacy in making any such presuppositions. Kripke, by contrast, is assuming the conclusion of his argument at a stage in the dialectic when it has not yet been established: to respond to an objection which, by his own lights, if successful means his conclusion never *will* be established.

McTaggart, I think, is engaged in the same fallacy as Kripke: he is inappropriately assuming his conclusion at a stage in the dialectic prior to when it is established. McTaggart charges the claim that things have objective A-properties with inconsistency. His opponent responds that his charge rests on a mistake, since the inconsistency relies on the A-properties being co-instantiated, whereas in fact they are merely had *successively*. McTaggart replies that to rely on succession is to presuppose that things have A-properties, and hence his opponent's response is circular. But *by McTaggart's own lights* the reason for rejecting the claim that things have A-properties is that this leads to inconsistency. So *if* it can be shown that no such inconsistency is generated, then by his own lights McTaggart should retract his claim that the postulation of A-properties is problematic. His opponent is attempting to show that, indeed, no such inconsistency is generated. At this stage in the dialectic, then, the inconsistency in the postulation of A-properties has yet to be established. It is *only* established if this attempt to demonstrate consistency fails. In which case, McTaggart cannot rely on the inconsistency of the A-properties to argue that this attempt does fail. That would be exactly analogous to Kripke's relying on meaning skepticism when responding to the simplicity objection: in each case, the conclusion is being appealed to at a point in the dialectic prior to its being established, and that is fallacious.

McTaggart's charge of circularity, then, is unconvincing. There is nothing dialectically inappropriate in McTaggart's opponent appealing to the A-properties of things when attempting to demonstrate the consistency of the claim that things have A-properties. Rather, the mistake is McTaggart's in objecting to his opponents' appeal to the A-properties before he has made the case that such properties lead to inconsistency.

Let us turn, then, to McTaggart's second objection to his opponents' defense of the A-properties: namely, that it leads to infinite regress. To briefly recap: McTaggart says the A-properties are inconsistent because they are incompatible, and yet everything has them all, since what is past was present, etc. His opponent says: there is no inconsistency, because incompatible A-properties are never had by the same thing *at the same time*, but only successively. Some thing is past and *was* present, but nothing is ever past and present at once. And there is no problem, says McTaggart's opponent, in incompatible properties being had by the one thing at different times. McTaggart responds:

> The difficulty may be put in another way, in which the fallacy will exhibit itself rather as a vicious infinite series than as a vicious circle. If we avoid the incompatibility of the three characteristics by asserting that M is present, has been future, and will be past, we are constructing a second A series, within which the first falls, in the same way in which events fall within the first. It may be doubted whether any intelligible meaning can be given to the assertion that time is in time. But, in any case, the second A series will suffer from the same difficulty as the first, which can only be removed by placing it inside a third A series. The same principle will place the third inside a fourth, and so on without end. You can never get rid of the contradiction, for, by the act of removing it from what is to be explained, you produce it over again in the explanation. And so the explanation is invalid.[16]

This is a baffling passage in many ways. McTaggart's talk of the "second A series," etc., is very misleading. As Ned Markosian points out, a series is just a collection of things.[17] The A-series is that collection of things (if any) that instantiate A-properties, and the B-series is that collection of things (if any) that are ordered by B-relations. But assuming that there are both A-properties and B-relations, *those are the same things*: the things that are earlier and later than one another, etc., are the same things that are present or past, etc. So the A-series just *is* the B-series. The A-properties and B-relations are distinct, but there is just one series of things that instantiate those properties and stand in those relations. Likewise, this "second A series, within which the first falls" is still just a series of the very same things—times, and events in times—which are now being said to have not only the first-order A-properties of being past, present, and future but also the second-order properties of having been past, going to be past, etc. But since this "second A series" is still a series of the very same things, it is not really a new series: it is the same A-series we already had. So McTaggart's suggestion here that we have an infinity of series is simply mistaken: we have *one* series of things, although perhaps we have to ascribe infinitely many properties, of increasing complexity, to the things in that series, if we want to ascribe any A-properties to them at all.

I think it best to ignore McTaggart's suggestion that the regress consists in the generation of a second, third, etc., A-*series*. I shall follow many a commentator in presenting the regress as arising because at each stage when one attempts to explain how incompatible properties can be had by the same thing, one appeals to their being had successively, which simply reintroduces the question.

[16] McTaggart (1908, p469). [17] Markosian (2004).

McTaggart presents us with a potential contradiction: M is present, will be past, and was future. So in *some* sense at least, each of the A-properties *is past, is present*, and *is future* are being attributed to M. How can this be, given that they are incompatible? Proposed answer: their incompatibility rules out only their being had *simultaneously*, but M only has these properties *successively*. Question: what does that mean? Answer: it means that the only property M *now* has is being present, but that in the past it had the property of being future and in the future will have the property of being past. We can rephrase this last claim in terms of the second-order A-properties M has: it is *now now*, it is *past future*, and it is *future past*. However, the problem that arose for the first-order A-properties now arises for the second-order ones. M is now now but it will be now past, so in *some* sense both the second-order A-properties *is now now* and *is now past* are being attributed to M, and yet these properties are incompatible: nothing can be both now now and now past. And of course, as before, we can attempt to resolve the incompatibility by claiming that all that is ruled out by the incompatibility of these second-order A-properties is their being had together *simultaneously*, whereas all we get is that they are had *successively*: M *now* has the property of being now now and *will have* the property of being now past. But again: we can rephrase this claim concerning the succession of second-order A-properties in terms of the having of third-order A-properties, and the same question will arise, and so on ad infinitum. At each stage, there is a set of A-properties all of which get ascribed in *some* sense to M but at least two of which are incompatible; the incompatibility is resolved if we can make sense of the incompatible properties being had by M merely *successively*, but this is merely to invoke the A-properties at the next level up, which gives rise to the very same question, and so we are off on an infinite regress.

McTaggart is clearly correct that there is an infinite regress. The question is whether it is vicious. I think it is not. I want to distinguish two types of infinite regress:

> *Benign regress of explanation*: At each stage, n, an instance of a puzzle is posed and an explanation offered, and this explanation gives rise to a new instance of the puzzle, forming stage n+1; but the success of the explanation at stage n does not depend on the puzzle at stage n+1 being resolved.

> *Vicious regress of explanation*: At each stage, n, an instance of a puzzle is posed and an explanation offered, and this explanation gives rise to a new instance of the puzzle, forming stage n+1; and furthermore, the success of the explanation at stage n depends on the puzzle at stage n+1 being resolved.

I think there is nothing especially problematic about benign regresses of explanation, but that there is something problematic about vicious regresses of explanation. (The hint is in the names.) In a benign regress, all that is happening is that the world presents you with infinitely many puzzles, where solving one puzzle points to another. What's wrong with that? Every puzzle gets solved, it is simply that answering one question raises the next. In a vicious regress, however, no puzzle ever gets solved, because the success of each solution relies crucially on the solving of the next puzzle, and so there is always a promise of a solution, but the promise is always postponed and never realized.

Here are examples of—what I take to be, at least—each kind of regress. Let us look first at what I take to be a benign regress. Suppose I believe in propositions, but that I want to claim that they are derivative beings: their existence is grounded in something non-propositional. In particular, I hold that the proposition p exists in virtue of it being possible that someone entertain the content that p. Furthermore, I analyze *in virtue of* as a relation between propositions: if p is true in virtue of q that is a matter of the *in virtue of* relation holding between the propositions p and q. Clearly, this view commits one to an infinite regress. For take any proposition, p. p exists in virtue of it being possible for someone to entertain the content that p. On this view, that is a matter of the *in virtue of* relation holding between two propositions, one of which is that p exists, the other of which is that it is possible for someone to entertain the content that p. Call this latter proposition "q." q exists in virtue of it being possible for someone to entertain the content that q. That means that the *in virtue of* relation holds between the proposition that q exists and the proposition that it is possible for someone to entertain the content that q. Call this latter proposition "r." And so on, ad infinitum. The regress is infinite but, I claim, benign. All that is happening is that each explanation generates a new question. But there is no sense in which the questions are not satisfactorily getting answered, because there is no sense that the success of the answers being given at each stage relies on there being an answer to the next question that is generated. The proposition that p exists is true in virtue of the proposition that it is possible to entertain the content that p. That can be a perfectly good explanation for why p exists. Giving this explanation raises a new question—why does this *other* proposition exist?—but the success of the explanation for why p exists does not hang on whether this new question is satisfactorily answered. There being an explanation for why there is this new proposition is not a requirement for the success of our explanation for why p exists. Even if this new proposition is a fundamental existent—one whose existence is a brute fact, not admissible of explanation—we would still have a good explanation of p's existence. We have already explained why there is the proposition p, now we can move on to explain why there is this other proposition. And so on: at each stage the question is satisfactorily answered. A new question is always generated, but why should we worry about that given that each receives a satisfactory answer?

By contrast, here is a view that I think generates a vicious regress. The view is a variant on Plato's Third Man, although different in detail from the standard presentation, and I am not attempting to attribute a view to Plato here. Suppose I am wondering how it is that things have qualitative natures—why is it that things are a certain way? I propose an answer: for a thing to be a certain way is for it to participate in a Form. So if we have some thing, A, that is some way, F, that is because A participates in some Form F. But it cannot just be that A's participating in *some* Form or other explains why A is F, for we have to be able to explain the *different* ways things are. There must be *many* Forms, and it must be how those Forms are that results in the things that participate in them being as they are. So A's participating in F is not the end of the story: to explain why A is F, F must have a certain character—*it* must be a certain way. But what we started off puzzled about was how things could be a certain way, so we cannot simply appeal to F being a certain way, we must explain *how* it can be that way. If things are the way they are by way of participating in Forms then F is the way it is by way of participating in at least one Form G. But as before, F's merely participating in *a* Form does not explain why it is the *particular* way it is; there must be something about the way G is that makes F's participation in it result in it being the way it is. So why is G the way it is? It must participate in at least one Form... and so on, ad infinitum. And here, I think, the regress is vicious. It is not simply that a new question is generated with each answer, it is that the answer is always postponed. We never learn why *anything* is the way it is, because the explanation given is only successful on the presupposition that the Form invoked is a certain way, and the explanation for it being that way is only successful on the presupposition that the next Form invoked is a certain way, and so on. At each stage, the explanation given relies for its success on the next puzzle being solved, but since each offered solution generates a new puzzle, the promise of explanation is always postponed and never fulfilled. We try to explain why A has a certain character and appeal to its participating in F; but our explanation is incomplete—A's participating in F only explains its character if F has a certain character. So all we have done is show that *if* F has a certain character, then A has a certain character. To complete the explanation, we need to show that F *does* have the required character, and so we appeal to its participation in G. But of course, all we have shown is that *if* G has the required character then F does too, and hence A does. No explanation is ever completed, and that makes for a vicious regress. This is completely different to the previous case, where each of the explanations given for why a proposition existed raised a new demand for explanation, but the first explanation does not depend for its success on the new demand being met.

The regress McTaggart identifies, I contend, is like the benign regress in the grounding of propositions, not like the vicious Third Man regress. A new demand for explanation is generated at each stage, but at each stage an explanation is given, and its success does not depend on the new demand for explanation being met. Thus there is no postponing of the explanation given at one level to the success of the explanation given at the next—a postponement that never ends if there are infinitely many levels. All we have in the McTaggart regress is an infinite sequence of questions, and an infinite sequence of satisfactory answers to each of those questions, and so the regress is benign and not vicious.

We start with the first-order A-properties, *is past, is present, is future*. We notice that in *some* sense we attribute each of them to M, since M is present, was future, and will be past. Puzzle: how can we attribute each of them to M when they are incompatible? Answer: we are merely attributing them to M successively. This is a perfectly satisfactory answer. That is all it takes to explain the sense in which incompatible properties can be attributed to M. There is nothing else on which the success of this explanation depends. A thing's participating in a Form is not a complete explanation of its character, for it is only A's participating in a Form of a certain character that explains its character, which is why the explanation gets postponed to the next level; but there is nothing that needs to be added to our explanation of the attribution of incompatible properties to M other than that they are merely had successively and never simultaneously.

Of course, *saying* that the first-order A-properties are had successively introduces a new question, for in saying that M has *is present* now but merely *had* the property *is future* we make salient the second-order A-properties *is now now, is past future*, etc. And the same question can now be asked of these properties. But this is simply a *new* question about a completely *different* set of properties. It is just like the propositions case when in explaining why the proposition p exists you make salient a new proposition (the proposition that describes the grounds for the proposition that p exists) about which you can then ask the same question: why does *it* exist? You get a new question about a new thing, but the fact that this new question is as yet unanswered does not undermine the answer you just gave to the previous question, because the success of that answer did not depend on this new question being satisfactorily answered. So the McTaggart regress is benign: we have an infinite sequence of puzzles and explanations, but each explanation is successful on its own, so nothing goes unsolved.

So neither McTaggart's charge of circularity nor of regress has led us to doubt the success of the obvious response to his charge that the postulation of A-properties leads to an inconsistency due to incompatible properties being had by the one bearer: the incompatible properties are only ever had successively, and that resolves that problem. However, there *is*, I think, a real problem extractable from McTaggart: a problem that relies on neither circularity nor regress but does require a further assumption about the metaphysics. The assumption is this:

Past Record: If something was the case, then it *is* the case in the past.

Now, clearly the *presentist* is not going to accept this principle: nothing *is* the case in the past, thinks the presentist, for the past is unreal. Things *were* the case—end of story. But on the face of it, the moving spotlighter and growing blocker ought to accept *Past Record* (and will be divided on whether to accept its future counterpart). According to the moving spotlight and the growing block views, the past is real. That is to say, seemingly, that what happened in the past remains a part of reality. Because there were dinosaurs, there *are* dinosaurs—where the "are" there is atemporal, describing how reality is across time. *Where* in reality are those dinosaurs? In the past, of course. So provided the "is" in the consequent of *Past Record* is an atemporal "is"—something *is* the case in reality as a whole, as it is across time—then it looks, *prima facie*, like a principle the believer in the past, like the moving spotlighter or growing blocker, should accept. Indeed, it looks like it is *because* they accept *Past Record* that these theorists can lay claim to some of the benefits they appear to have over presentism: in particular, those concerning truthmakers for historical truths. The growing blocker and moving spotlighter are supposed to have an advantage over the presentist in saying what it is that makes true claims about what happened in the past. What makes it true that there were dinosaurs? The dinosaurs that exist in the past make it true. What makes it true that Frank the T-Rex was brown? Frank the T-Rex's being brown in the past makes that true. This is meant to be an advantage over the presentist who, lacking this past ontology, simply has to take it as a brute fact that things were that way (or postulate the present with surprising ontology that settles these past truths, thus undercutting their claim to having an ontology that is theoretically parsimonious with respect to their rivals). But this advantage only straightforwardly accrues to the non-presentist theories if they accept *Past Record*. For suppose *Past Record* is false. Suppose that the past exists but that in becoming past it changes character somehow, so that how things were is not how the past *is*. Just for the sake of example, suppose things become colorless when they become past. What makes it true on this metaphysics that Frank the T-Rex was brown?

Not Frank the T-Rex's being brown in the past, for Frank the T-Rex is *not* brown in the past, he is colorless. Frank the T-Rex's being brown is nowhere to be found in reality: that is a way things were, but it is not a way anything is, even atemporally speaking. So what makes it the case that Frank the T-Rex was brown? It looks like we have to admit a brute tensed fact to the effect that actually colorless Frank *was* brown. But then, what advantage is it to having Frank around at all? If there must be brute tensed facts concerning how he was, why not simply go all the way to presentism and admit brute tensed facts concerning there *having been* Frank and him *having been* brown. Once you have got to have *some* brute tensed facts, why not just go all the way? In which case, no need to admit the reality of the past in any sense at all.

So *Past Record* does, *prima facie*, look like an attractive principle for the non-presentist A-Theorist to adopt. However, adoption of this principle quickly gets us into McTaggart-esque trouble. For one way things were is that M, which is now present, was future. So M's being future is a way things were; M's being future is something that was the case. So by *Past Record*, M's being future *is* (atemporal "is") the case, in the past. But M is now present, and how things are now is also a part of how things atemporally are across time. So both M's being present and M's being future are the case, atemporally. So now ask the question: what is M like? Not: what is M like *now*? What is M like, *simpliciter*? It looks like there is pressure to say both that M is present simpliciter and that M is future simpliciter, since both M's being present and M's being future are the case, atemporally speaking. To say merely that M is present simpliciter is to ignore the sense in which M's being future is a part of how things are, and likewise *mutatis mutandis* if we say merely that M is future simpliciter. It looks like we have to say both things, to respect both the reality of how things are now and the reality of how things were. And yet we *cannot* say them both, for surely nothing is both present and future simpliciter.

This version of McTaggart's argument is usefully compared to the problem of temporary intrinsics.[18] I am 6ft tall *now*, but I was 5ft tall. But what height *am* I? A natural first reaction is to think that there is simply no puzzle: I have already said what height I am, I am 6ft tall. I *was* a different height, but that is not what height I *am*. But that misses the point: the question is not asking what height I am *now*, but asking what height I am *simpliciter*. Compare: I can ask about whether there are dinosaurs. And I am not asking about whether there are dinosaurs *now*, or whether there are or have been or will be dinosaurs *at some time*, for everyone agrees that there are no dinosaurs now but that there were. I am asking whether

[18] Cf. Craig (1998).

there are dinosaurs, simpliciter: and in response to this question about how reality is simpliciter, not at a time, the presentist will answer that there are no dinosaurs whilst the eternalist will answer that there are. And just as I can ask how reality is simpliciter with respect to containing dinosaurs, so can I ask how it is simpliciter with respect to my height. And *prima facie*, it is an inconsistent way: it is both the case that I am 6ft and 5ft tall, since here I am in reality being 6ft tall (at the present time), but there I am in reality being 5ft tall (at the past time).

There are two ways of resolving the apparent inconsistency generated by temporary intrinsics: we can deny that when we predicate heights to a thing we are thereby attributing to that thing a monadic property that is had simpliciter, or we can deny that the things that have height properties simpliciter ever appear in reality with an incompatible height property. On the former strategy, some claim that apparent temporary intrinsic properties are actually relations to times.[19] So nothing is 6ft simpliciter, any more than anything is *taller than* simpliciter. You cannot be taller than, simpliciter, you can only be taller than some thing—similarly, on this view, you cannot be 6ft simpliciter, you can only bear the *being 6ft at* relation to some time. The question "What height am I?" is a good one to ask, but only because there is a salient time—the time of utterance—such that the correct answer is the height relation I bear to that time. But if I ask what height I am *simpliciter*, not *at* any time, the question betrays a misunderstanding: the best answer one can give is that I bear such-and-such a height relation to time t, so-and-so a height relation to time t*, and so on. And these facts never change, and they are not incompatible, so we avoid the apparent inconsistency. Others taking this general strategy hold on to the claim that *being 6ft* is a monadic property rather than a relation, but deny that such properties are had simpliciter: rather, they can only be had in a certain way, where for each time, t, there is a way of having those properties t-ly.[20] And so I have *being 6ft* 2015-ly, but I lack it 1980-ly. And those two claims never change, and are compatible, so again I avoid contradiction.[21] On the second strategy, *being 6ft* is a monadic property and is had simpliciter, but the things that have such properties do not appear anywhere in reality with any incompatible property. On one view, that is because the things that bear those properties are not persisting objects but rather temporal parts of persisting objects, which are located only at an instant.[22] I am 6ft tall now and was 5ft tall in the past, but if you ask how tall I am

[19] See van Inwagen (1990). [20] See Haslanger (1989).

[21] Although I will argue in §4.5 that these solutions, while they avoid the inconsistency, fail for other reasons.

[22] See Lewis (1986), Sider (2003a), Hawley (2001).

simpliciter: well, I am not the kind of thing that has those properties simpliciter—there is a temporal part of me that exists now and is 6ft tall simpliciter, and another temporal part that exists in the past and is 5ft tall simpliciter. But since those temporal parts of me are distinct things, there is no contradiction.

Now return to the McTaggartian puzzle. Just as I can ask what height I am simpliciter, so can I ask whether Caesar's crossing the Rubicon is past simpliciter. Not past *then*, not past *now*, just whether it is past. And surely the A-Theorist must answer that it *is* past simpliciter. It is before the objective now, thus it is past—end of story. But it *was* present, so by *Past Record* it must *be* present in the past. But the only place it *is* is in the past, so it must be present, simpliciter, as well. So it is past simpliciter, and present simpliciter. But those are incompatible properties!

Are any of the above ways out of the problem of temporary intrinsics ways out of the McTaggartian paradox? No analog of the temporal parts solution, surely: it is not like there is a part of Caesar's crossing the Rubicon that is past and a part that is present. All of the event is in the past, so it should all be past simpliciter; and the problem arises because there is also reason to say that all of it is present simpliciter (since each part of it *was* present and hence, by *Past Record, is* present in the past). Perhaps, then, the A-properties are disguised relations to times, or are properties that are had not simpliciter but only ever in a certain way—t-ly, t*-ly, etc? No, for that simply abandons the A-Theorist's guiding thought: that there is a unique privileged time. *Every* time would bear the *is present* relation to itself, and *every* time t would be present t-ly: there would be nothing left to the claim that one time is *uniquely and objectively special* in being present.[23]

Now of course, the presentist can avoid both this McTaggartian puzzle and the problem of temporary intrinsics.[24] The presentist agrees with the temporal parts theorist that the way to respond to the problem of temporary intrinsics is to deny that the same thing is to be found in reality having both the property *is 6ft* and the property *is 5ft*. But that is not because the bearers of such properties are instantaneous beings: rather, it is because reality only ever encompasses a single moment of time—that which is present. So I am now 6ft tall, and this is not a matter of me bearing some relation to the present time, or a matter of my having this property a certain way: I have this property simpliciter. That is how I am *now*, but it is also how I am *simpliciter*, for how things are simpliciter *just is* how

[23] Taking such a route leads us towards Fine's non-standard realisms about tense. See §2.4.

[24] See Zimmerman (1996) for the presentist response to the problem of temporary intrinsics. Craig (1998) holds that McTaggart's argument poses a problem for non-presentist A-Theories, but not for presentism.

things are now, since present reality is the extent of reality. I was 5ft tall, but nowhere in reality am I 5ft tall, for how things were is no part of how reality *is*, for the past is not real on this metaphysic. So the presentist resists the move from "Things were p" to "There is a past time at which p." My being 5ft is no part of reality, nor is the presentness of Caesar's crossing the Rubicon. I am 6ft simpliciter, and Caesar's crossing the Rubicon is past simpliciter. Things were different—I *was* 5ft simpliciter and Caesar's crossing the Rubicon *was* present simpliciter—but their being so is no part of how things are, and thus we are not forced to ascribe incompatible properties to things. The problem only arises once we grant that how things were (or will be) is nonetheless a part of how reality *is*.

2.2 Smith's McTaggart

So on the face of it, the McTaggartian puzzle is a serious problem for non-presentist A-Theories, at least if they accept *Past Record*, but no problem at all for presentism; thus we have a *pro tanto* argument for the claim that the A-Theorist ought to embrace a presentist ontology. We will return to this issue in the next section, but before we do let us look at a recent attempt by Nicholas J. J. Smith at putting forward a McTaggart-esque argument against *all* versions of the A-Theory.[25] It will prove worth taking this diversion now, as we will learn some lessons from answering Smith's problem that will help us address the above problem.

The presentist, even though they hold that Caesar's crossing the Rubicon is no part of reality, of course has to hold that it *was* the case that Caesar's crossing the Rubicon is part of reality. The growing blocker and the moving spotlighter get into trouble, if they accept *Past Record*, because they then have to hold that Caesar's crossing the Rubicon is present in the past. But the presentist says no such thing: she merely holds that it was the case that Caesar's crossing the Rubicon is present. And for the presentist, this tense operator—"It was the case that..."—is primitive: in particular, she denies the claim that its truth conditions are given by "It is the case in the past that..." And this is how she avoids the problem for the non-presentist A-Theorist that we considered at the end of the last section.

Smith argues that this does not solve the problem. He says the primitiveness or otherwise of the operator is neither here nor there: what matters for the purposes of getting the problem going is that the operator is "detachable,"[26] in the sense that using these tense operators gives us information for how things were/will be. So suppose I say "In a hundred years, there will be lunar colonies." If this is true

[25] Smith (2011). [26] The terminology is Smith's.

and if the tense operator "In a hundred years..." is detachable then it follows that if I wait a hundred years there will be lunar colonies. For another example, consider the following argument:

1. By the time you get to the conclusion of this argument, it will be the case that p.
2. Therefore, p.

If the tense operator is detachable, this argument is necessarily truth-preserving in the sense that you will never be able to utter it and have uttered a true premise but a false conclusion. The detachability of the tense operator means that (1) gives you information about how reality will be if you just wait long enough: "long enough" here is the time it takes you to get to the conclusion, which says that reality is the way you have been waiting for. So if (1) is true when you utter it, so must (2) be when you utter it.

One could believe in tense operators that are not detachable. I could tell you now that elephants are about to appear in the room. Now wait for it... And now you challenge me on the absence of elephants, and I agree that there are not now any elephants in the room but I refuse to retract my earlier claim, saying that I was only making a claim about how things will be, and that this was true despite things not being that way now (after we have waited the allotted time). There is nothing inconsistent about such operators. But Smith is surely right that genuine tense operators must be detachable: if its being true that it will in X years be that p does not let me wait X years and then truly conclude that p, then we have lost all grip of the subject matter.

But if tense operators are detachable and thus give us information about how things were/will be, then the presentist should be able to join the eternalist in giving us a spacetime diagram showing things being different ways at different times that accurately represents reality as it is, was, and will be. Sure, the eternalist thinks that every part of the diagram represents something *real* whereas the presentist will deny this, but both should agree that it gives us a perspicuous representation of reality: the eternalist because it tells us how reality is at the different times that there are, the presentist because it tells us how reality is at the time that is real, and also how it was/will be at the times that are not but were/will be real.

But how is the A-Theorist, whether presentist or otherwise, to draw this diagram? It needs to tell us, for starters, that the one unique privileged time is present. Let us represent that by coloring one vertical slice of the diagram green. And it needs to tell us that the other times—the ones that are not real according to the presentist, or not privileged according to the non-presentist—are not present: so we should not color them in green. What should we draw when drawing those past and future times? We should draw them as they were/will be. If there were dinosaurs at t, then at the point of the diagram which represents t, we should draw some dinosaurs. If it was the case that p, then we should represent things being p at some point on the diagram before the bit that represents the present. But the past times *were* present. So we should represent them as being present before the present: so at the point on the diagram which represents the past time t, we should represent t as being present, by coloring it in green. But we have already said that we have *not* to color this time in green, since it is past. Our instructions are contradictory! But these instructions should be possible to follow given the detachability of the tense operators, says Smith: so the problem is not with the instructions, but with what we are trying to represent. The A-Theory—even a presentist version—is inconsistent, as witnessed by the impossibility of representing how reality would be were it true.

Here's the essence of Smith's argument, as I see it, as pressed against the presentist. The A-Theorist, as we saw at the end of the last section, gets into trouble if she accepts the move from "It was/will be the case that p" to "It is the case in the past/future that p." The presentist denies that inference. But if her tense operators are detachable, as they must be, then she must accept the move from "It was/will be the case that p" to "For any representation that correctly and fully represents how reality is, was, and will be, that representation represents a past/future time at which p." And that is enough to generate the puzzle because now we must represent each past/future time as being non-present, since they *are* non-present, but we also must represent them as being present, since they were/will be present. And that is incompatible with this being a correct representation of reality, since in reality no time is both present and non-present. So there is no way to adequately represent reality as the presentist sees it, says Smith: and that is because, he thinks, their metaphysic is, like other A-Theories, inconsistent.

I think that Smith's argument fails. Before we see why, let us note that suspicion should be raised by the fact that if Smith's argument is sound then it also apparently rules out what is *by far* the most popular view on modality. The modal equivalent of the claim that there is a privileged time—the present—is the claim that there is a privileged world—actuality; the modal equivalent of the claim that there are non-present times in reality is the claim that there are non-actual worlds in reality. And so the view analogous to B-Theoretic eternalism—that there are non-present times but no privileged present—is Lewis's modal realism:[27] there are non-actual worlds, but no privileged actuality. The view analogous to A-Theoretic non-presentism—that there are non-present times

[27] Lewis (1986).

but one objectively privileged time (the present)—is (with a caveat to which we will come back in the next section) the view championed by Phillip Bricker: that there are non-actual worlds, but one objectively privileged world (actuality).[28] And the view analogous to presentism—that there is a privileged time (the present) and it is the only time there is—is the view that almost everyone believes: that there is a privileged world (actuality) and it is the only world there is.[29]

Bricker's view, on the face of it, faces a serious McTaggart-esque challenge. He believes in many worlds, but only ours has the special property of being actual. But the non-actual worlds, of course, *could* be actual—this is the claim that is analogous to the A-Theorist's claim that the non-present times were/will be present. But on this metaphysic, what *could be* the case *is* the case at some world. So take some non-actual world w. It *could* be actual, so it *is* actual at some world in modal space. At what world is w actual at? The only candidate world is w itself. So w is actual *at w*. But on this metaphysic, for something to be true *at w* is for w to be that way: there are dinosaurs *at w* iff w contains dinosaurs. So if w is actual *at w*, w is actual. But w is not actual. So w is both actual and not actual. Contradiction!

Now, Bricker thinks this problem can be solved, but we will come back to that in the next section, when we will ask whether he offers solace to the non-presentist A-Theorist. For now, note that the Lewisian, of course, faces no puzzle, for on their view "is actual" does not predicate some special metaphysical property: "actual" is an indexical, and every world is actual from its own perspective. And *prima facie* the actualist faces no puzzle for although they think that "is actual" predicates a monadic property of worlds, they think that only our world has it, and while that property *could* be had by other worlds, they resist the move from this to the conclusion that the property *is* had by other worlds at other worlds, since they do not believe in other worlds and hence resist in general the move from "things could be that p" to "there is a world at which p."

Of course, the actualist believes in only one *genuine* world, but she may believe in many abstract *representations* of other worlds. But *prima facie*, this does not invite the same problem, since the ersatzer will not accept that if something could

[28] See Bricker (2006), *inter alia*. In fact, to accommodate the possibility of island universes, Bricker allows that there may be more than one actual world, but I will ignore this complication going forward: nothing is going to hang on this simplification.

[29] Of course, many actualists will take possible worlds talk seriously in some sense, in that they believe in *ersatz* worlds. But while such actualists will say things like "I believe in many worlds," in the relevant sense of "world" they believe in only one. Ersatz worlds are not really worlds (hence the term "ersatz"): they are abstract *representations of worlds*. Similarly, some presentists believe in many *ersatz* times (Bourne 2006): but that is compatible with the presentist's thesis that there is only one time, because an ersatz time is not really a time, it is merely a *representation* of a time.

be the case then there is an ersatz world that is that way. Rather, they will accept only that if something could be the case then there is an ersatz world (i.e. an abstract representation of a world) that *represents* things being that way. And there is no contradiction, seemingly, in w being non-actual, simpliciter, but there being a *representation* of w that *represents* w as being actual, simpliciter.

But if Smith's argument against presentism is good, then this response on behalf of the actualist is too quick. Sure, the actualist, unlike both Lewis and Bricker, thinks that reality only encompasses the actual. But they should still be able to draw a worlds diagram, showing how reality is at different worlds: it is just that while Lewis and Bricker will take such a diagram to represent reality by representing what there is and how it is, the actualist will take such a diagram to represent reality by representing how things are and also how they are not but could be. She will not believe in the reality of the things represented on the diagram that are not actual, but she will believe that they represent reality in that how these bits of the diagram represent things as being represents a way reality could have been. So how should she draw this diagram? She should color the part that represents the actual world green to represent its being actual. She should not color the parts that represent non-actual worlds green, so as to represent the fact that they are not actual. She should draw the non-actual worlds as they would be were they actual, so that they represent correctly how things could be. So she should draw some talking donkeys at the part of the diagram that represents a way things could be such that there are talking donkeys. And, in general, if things would be such that p were w actual, she should draw the part of the diagram that represents w as being such that p. So since the non-actual world w *could* be actual, she should color it green to represent its being actual *at w*. But we already said she should not color it green, to represent its being not-actual. So she should both color it green and not color it green. The instructions are inconsistent. And if Smith drew the right conclusion about the temporal case, then it seems that the analogous one must be drawn here: the inability to follow the instructions in constructing the representation reflects the underlying inconsistency in the metaphysic. Just as it is inconsistent for there to be a privileged present, even if the present time is the only time there is, so is it inconsistent for there to be a privileged actuality, even if the actual world is the only world there is.

Headline news! Many people might be able to live without the A-Theory, but to be forced toward Lewisian modal realism is going to be a hard pill for the vast majority to swallow. Thankfully, I think Smith's argument does not work, and both presentism and actualism are viable metaphysics.

All that Smith's argument shows, I think, is that there are limits to how you can accurately represent how reality is according to presentism and actualism.

In particular, the following rules cannot be consistently followed given the truth of these metaphysics:

(R1): Represent all the times [/worlds] that there are, were or will be [/are or could be].

(R2): Choose a representation scheme that matches every way for a time [/world] to be to a feature of the part of the representation that represents that time [/world].

(R3): If a time t [/world w] is F, make the part of the representation that represents t [/w] represent t's being F [w's being F], in accordance with (R2).

(R4): If a time t [/world w] was or will be F [/could be F], make the part of the representation that represents t [/w] represent t's being F [w's being F], in accordance with (R2).[30]

Rules (R1)–(R4) cannot consistently be followed if how reality as a whole is changes or could be different, for then (R3) will tell you to make the representation one way, to represent how reality as a whole *is*, and (R4) will then tell you to make it an incompatible way, to represent how reality as a whole *was* or *will be*, or how it *could have been*. In particular, suppose the representation scheme we choose when complying with (R2) tells us to match the temporary [/accidental] property of something's being present [/actual] with the representation of that thing on the diagram being green, and a thing's not having that property with its representation not being green. Then (R3) tells us to make only the representation of the present time [/actual world] green, but (R4) tells us to make the representation of every time [/world] green.

But this is only a limitation on how we can *represent* the metaphysics; it reflects no underlying incoherence in the metaphysics itself. Or at least, I see no reason to think that it does; and Smith, I think, fails to make that case. After all, it is not like

[30] Example. Here is a way for a time to be: *containing sea-battles*. So what do the rules tell you to do? First, following (R2), match that feature that times can have to a feature that a part of the representation can have: what features you can choose from here depends on what the representation is like, but let's assume parts of it can be colored, and let's choose to map *containing sea-battles* to *being red*. Now, following (R3), we will make a part of the representation red if and only if that part represents a time that in fact contains sea-battles. If you are a presentist, the only time that can contain sea-battles is the present (since it's the only time that is real): so we should color the part that represents the present red iff there is now a sea-battle. The non-presentist, however, will also need to color the part that represents 1805 (e.g.) red, since there *was* a sea-battle in 1805 (the battle of Trafalgar), and the non-presentist thinks this means that there *is* a sea-battle at this past time. But even though the presentist does not color the part that represents 1805 red at this stage, they will at the next stage. For, following (R4), the presentist must color that part red in order to represent that there *was* a sea-battle at this time *when* that time was present. Continue for every way a time can be.

reality is in principle unrepresentable given these metaphysics: it is merely that it is unrepresentable *in a certain manner*, namely, by following (R1)–(R4). But I do not see why this should be a problem: it is unsurprising that there is such a limit on representation. We are being told to represent all the times [/worlds] that were or will be [/could be] present [/actual], and we are being told to represent how reality is, so we need to represent the non-present [/actual] times [/worlds] as being non-present [/non-actual]. But we are also being told to represent how reality was or will be [/could be] *using the same system of representation*: that is, such that if we are representing the present time [/actual world] as being F by making the part of the diagram that represents this time [/world] F*, then we should represent the fact that things are F *at a time* [/*at a world*] by making the part of the diagram that represents that time [/world] F*. Of course this simply cannot be done if how reality is changes or could be different! We need to be able to represent the non-present times [/non-actual worlds] as being non-present [/non-actual], but we also need to represent that they were or will be [/could be] present [/actual]—hence that they are present [/actual] *at themselves*. So the representation of reality needs to represent them being non-present [/non-actual] but present [/actual] at themselves. This means that we *cannot* represent it being *true at t* [/*at w*] that t [/w] is F by doing to the representation of t [/w] what we would do to simply represent t [/w] being F. We want to represent t's being present *at t* [/w's being actual *at w*], but we simply cannot do this by making the representation of t [/w] green, as we would do were we representing t's being present [/w's being actual] simpliciter. Rather, we need to represent presentness *at a time* [/actuality *at a world*] differently from how we represent presentness [/actuality] simpliciter. How we represent times and worlds as being present and actual must differ from how we represent them as being present and actual *at themselves*. That is certainly not surprising given presentism or actualism, since on those metaphysics how times and worlds *are* simpliciter is different from how they are *at* themselves, since non-present times [/non-actual worlds] are unreal, simpliciter, but they are real *at themselves* (which means just that they were or will be real [/would be real], when they were or will be present [/were they actual]). But it is not surprising, really, given *any* version of the A-Theory, or any non-Lewisian view of modality, since those metaphysics take how times or worlds to be simpliciter to be something that varies across time and across modal space. Any such view has to think that there is some feature that is had by a time/world but which is lacked by it *at that time/world*, which is just to say that, although the time/world in fact has that feature, it will not or would not have that feature when that time is present or were that world actual. And so any such view is going to have to represent a time or a world's having that feature differently from how it represents that time or world having that feature *at itself*.

No presentist or actualist should ever have expected to be able to follow (R1)–(R4), and it is no cost that they are not able to do so. In distinguishing between representing a time or world as being F and representing its being F at itself, they need to replace (R1)–(R4) with:

(R1*): Represent all the times [/worlds] that there are, were or will be [/are or could be].

(R2*): Choose a representation scheme that matches every way for a time [/world] to be to a feature of the part of the representation that represents that time [/world], and every way for a time [/world] to be *at a time* [/*at a world*] to a feature of the part of the representation that represents that time [/world].

(R3*): If a time t [/world w] is F, make the part of the representation that represents t [/w] represent t's being F [w's being F], in accordance with (R2).

(R4*): If a time t [/world w] was or will be F [/could be F], make the part of the representation that represents t [/w] represent t's being F at t [w's being F at w], in accordance with (R2).

These rules can be consistently followed. Suppose we are at the second moment of a three-moment history: so time t2 is present, t1 is past, and t3 is future. Suppose further that there are only two temporary ways for things to be in this world: redness can ensue, or blueness can ensue. Lastly, suppose that redness presently ensues, but that blueness did ensue and will ensue again. Now let us represent this world in accordance with (R1*)–(R4*). Let us represent the times t1–t3 by the names "t1"–"t3"; represent one time being before another by drawing the name that represents the former time to the left of the name that represents the latter time; represent blueness ensuing at a time by writing the name of that time in bold and represent redness ensuing at a time by writing the name of that time not in bold. Represent a time being present by underlining the name of that time. Now how to represent a time's being present *at that time*—that is, how to represent that a past time *was* present and that a future time *will be* present? We cannot underline them to represent this, as (R1)–(R4) would have us do; but no matter, let us just do something else: let us put a star next to the name of a time to represent that it is present *at that time*. Then we can completely and accurately represent this world as follows:

t1* t2* **t3***

Nothing is missing from this diagram: it tells you everything you need to know about reality. What it *does not* do is show you that things *will be* such-and-such in the same way you show things as being such-and-such, for things are represented as being present by being underlined, but things are not represented as being present in the past by being underlined in that part of the diagram that represents the past—rather, we use a different method of representation for this; we put a star next to the things that were, are, or will be present. So there is a lack of uniformity in how our representation represents: but who cares, so long as we can represent how reality is some way or other?

And so I think the presentist and actualist can rest easy. They face an essential limitation in how they can accurately and completely represent reality as they see it—a limitation that does not apply to the B-Theorist or the Lewisian modal realist (who can follow (R1)–(R4) without inconsistency). But I at least see no reason to think that this limitation on representation reflects any problem in the underlying metaphysics: reality as these metaphysics have it *can* be represented, just not as simply as otherwise.

Of course, questions remain unanswered in giving this representation. For example, what exactly is it we are representing in reality when we represent time t1 being present *at itself*? And how does that feature of reality connect to time t2 being present simpliciter? If *Past Record* were true we would have an answer: representing a past time being present at itself is to represent it as being present simpliciter in the past. But in abandoning *Past Record*, we give up on such an answer. So what in reality corresponds to putting a star next to the names of each time in the representation above? As yet, we have been given no answer to this. But that is okay: giving a representation of how reality is can leave it unsettled what the underlying metaphysical facts are that make that representation an accurate one. We want to know the answer to that question, of course, and I will attempt to answer it in chapter four, when I present my version of the moving spotlight metaphysic. For now, the point is not to say what the underlying metaphysic is that makes the A-Theoretic representation of reality true, but merely to show *contra* Smith that we can consistently give such a representation in the first place.

2.3 Non-presentist A-Theories and McTaggart

Let us take stock. Neither McTaggart's nor Smith's argument for the inconsistency of the A-Theory in general has proved convincing. There *is*, however, a *pro tanto* case for the inconsistency of non-presentist A-Theories that accept the principle *Past Record* (and/or its future counterpart)—and so, since there is some reason to think that the non-presentist *ought* to accept *Past Record*, we have here an argument that the A-Theorist ought to be a presentist.

What to make of this argument against non-presentist A-Theories? As I said in the previous section, the modal analog of non-presentist A-Theories is Phillip Bricker's view on which there are many genuine possible worlds but where, unlike on Lewis's theory, there is a unique objectively actual world. But of course, it is *contingent* that the actual world is actual: each of the possible worlds that are in fact not objectively actual *could* have been so. That is the claim that is analogous to the non-presentist A-Theorist's claim that times which are not in fact present were or will be so. And so we would expect there to be a modal analog of our McTaggartian argument against non-presentist A-Theories that threatens Bricker's non-standard modal realism. And indeed there is, and Bricker responds to it: so let us look at what Bricker says, and see if our non-presentist A-Theorist can find solace there. Suppose Bricker accepts the modal analog of *Past Record*:

> *Possible Record*: If something could have been the case, then it is the case in modal space.

Possible Record sure *sounds* like a good thing for a believer in genuine possible worlds to hold. Isn't the whole point of believing in genuine possible worlds that you then have a reality in which every possibility is realized somewhere or other, and hence have an ontological underpinning for truths concerning what could have been the case? If something could have been the case but is not the case anywhere in modal space, then what accounts for the fact that it *could* have been the case? Is it a brute modal fact? Wasn't the point of genuine possible worlds that we could do away with brute modal facts? But if *Possible Record* is true and there is absolute actuality, then it looks as if we can run the modal analog of the McTaggartian argument. For consider some non-actual world, w. w could have been actual. So by *Possible Record*, w is actual somewhere in modal space. Where in modal space? At w itself, surely; where else? So w is actual. But w is not actual. Contradiction.

Bricker thinks that this modal analog of the McTaggartian paradox can be solved, and to do so he invokes the distinction between what is true *at* a world and what is true *of* a world. What is true of a world is just how that world is. And only our world is actual, so *is actual* is true *of* no other world at all. But that is okay: what *could* be the case depends on what is true *at* some world. Most of the time, what is true at a world is what is true of it, thinks Bricker, but not in the case of *is actual*: this is true *at* every world, but true *of* only our world. The latter secures the absoluteness of actuality, the former secures the evident modal truth that which world is actual is a contingent matter, and that every world could be actual.

This is to reject *Possible Record*. "w is actual" is not true anywhere in modal space on Bricker's view. Look across the space of possible worlds and you will find that only one of them—our own—is actual. However, it is still taken to be true that w could have been actual, and the reason for this is that we do not demand, for that possibility to obtain, that w's being actual is realized somewhere in modal space: rather, we demand only that w *represent* itself as being actual, and allow that it can do so without in fact being actual. This is to borrow a trick from the ersatzer, who of course has to allow that ersatz worlds or times can represent something as being the case at themselves without in fact being that way. After all, those ersatz times and worlds are entirely abstract, but they must represent themselves as being such that various concrete happenings are going on at them, in order to account for the fact that various concrete happenings occurred/will occur when the genuine times that are represented by those ersatz times were/will be present, or that various concrete happenings would have occurred had the genuine worlds that are represented by those ersatz worlds been actual.

We must be able to explain *how* a world or time—whether ersatz or genuine—represents something to be true at them, despite their not being that way themselves. For some ersatzers, at least, this is easy: if ersatz worlds/times are sets of propositions, for example, for a proposition to be true at that world/time can be just for it to be a member of that world/time. It is not so easy for the believer in genuine worlds and times, however. *Real* times and worlds—such as the ones we are actually at now—are not paradigmatically representational entities. A world or time can represent things as being the way it in fact is, of course, just as I can represent there being a human by virtue of my being a human. And indeed, this is in general the account of representation the believer in genuine times and worlds will give: Lewis and Bricker hold that a world represents "There are talking donkeys" as true at it by having talking donkeys as parts, and the moving spotlighter (as traditionally understood, in any case) holds that a past time represents "Caesar crosses the Rubicon" as true at it by having Caesar's crossing of the Rubicon occurring at it. But how do these non-paradigmatically representational entities represent things as being the case at them when they are not themselves that way?

Indeed, one might worry that *allowing* them to do so undercuts the motivation for believing in such things in the first place. Wasn't the whole point of eternalism and modal realism to give an ontological ground of tensed and modal truth? So-and-so was the case *because* it *is* the case at some past time; such-and-such is possible *because* it *is* the case at some other world. *Prima facie*, then, the distinction between what is true *at* a time/world and what is true *of* a time/world has no place; what is true *at* a time/world—what was/would be the case were it present/actual—should simply be a matter of how that thing *is*, that is, what's true *of* it. The presentist and actualist, by contrast, are making no such claims about the grounds of tensed or modal claims: the facts about what was or will be the case, or could be the case, are brute—or at least, they do not hold in virtue of non-present times or non-actual worlds being a certain way.[31] And so there is no problem in distinguishing between how those times/worlds are and how they are *at themselves* when we are merely aiming to *represent* those brute tensed and modal facts. The problem only arises once we take tensed or modal truths to be a reflection of how reality *is* elsewhere along the temporal or modal dimension. For then, seemingly, other times only *will be* present, and other worlds only *could be* actual, if there are times and worlds elsewhere on those dimensions which *are* present and actual—which is the problematic conclusion, for we want to hold that there is a *unique* present time and actual world.

Of course, there is precedent for the believer in genuine possible worlds to allow that they can represent something's being true at them without in fact being that way. The case that must be springing to the reader's mind is Lewis on *de re* modality. "Obama is 4ft tall" is not true *of* any world, since no world but ours contains Obama, and Obama is not 4ft tall in this world; but it is true *at* other worlds, namely, those worlds at which there is a counterpart of Obama who is 4ft tall.[32] We have a story about how worlds represent Obama being a certain way even though Obama is not a part of those worlds: they represent Obama vicariously, by containing counterparts of him—things similar enough to him in relevant respects; and they represent Obama being F by having as a part a counterpart of Obama that is in fact F. So what we need in the case of *is actual*, if Bricker's view is to be acceptable, is an account of how worlds that are not actual represent themselves as being so. Bricker's answer is to "simply accept whatever account a Lewisian provides for representation by worlds, and then add a separate clause to deal with the property of actuality... *world w is actual* is true at *v* if and only if *w* is identical with *v*. The contingency of actuality then follows immediately: *for any world w, possibly, w is the one and only actual world.*"[33] The temporal analog of this, of course, is to hold that out of all the many genuine times there are, exactly one is present. For something to be true *now* is for it to *be* the case at that one time. *Most of the time*, for it to be true that things *were* a certain way is for them to be that way in the past. Except when we come to say that past times *were* present. Then it is just true automatically, for "t is present" is true at t* just when t=t*.

[31] Bourne (2006) is an exception; I do not think his truthmaking project has a chance of success, but I will have to argue for that another time.

[32] Lewis (1986, ch.4).

[33] Bricker (2006, pp52–3). Bricker actually considers two ways of having absolute actuality, and there is a different, slightly more complicated clause for the other theory, but the difference is not relevant for present purposes.

It is easy to find this dissatisfying. It looks ad hoc. What does their being self-identical have to do with a world's or time's being actual or present? Bricker's clause looks like it has simply been engineered to get the correct things coming out as true, but without having any independent justification. We have a good sense of *why* "There are talking donkeys" is a good thing to say about what is going on at some world, w: because it has talking donkeys as parts. Why is "w is actual" or "t is present" a good thing to say about what is going on at w or t, just because w/t is identical to itself? There is no independent reason to think that that is a good description of modal or temporal space; the only reason to accept Bricker's clause is because it yields the result we desire: that every world could have been actual, and every time was or will be present.

Here is a way of seeing the problem. Consider the "stuck spotlight" view: the past, present, and future all exist, the present time is objectively privileged, and it never changes that it is *that* time that is privileged. That seems to me like a coherent hypothesis about how things are. But even were things that way, nothing would stop us adopting Bricker's clause as a way of speaking. Even though we knew the spotlight never moved and that no time other than the actual present ever was or ever will be present, we can still talk *as if* past times were present and future times will be. Just say that a time is present at itself if and only if it is identical to itself. If we start to talk this way, we will start saying exactly the same things as the moving spotlighter. But we have not changed the metaphysics, we have just started using tensed language in a way that does not reflect the underlying metaphysics. So there is a worry that the moving spotlighter, if she adopted the Bricker proposal, could not distinguish herself from the stuck spotlighter with a fancy semantics. Likewise, *mutatis mutandis*, in the modal case: the worry is that Bricker cannot distinguish his view from the "stuck actuality" view: the view on which there is an objectively actual world, and it is necessarily that world that is actual, but where we have a fancy semantics whereby we can *say* that other worlds could have been actual.

But here, I think, a difference is revealed between the two views. This is not a damning problem for Bricker's view of modality, but it *is* a problem for the moving spotlighter who makes the analogous move. The difference arises because Bricker's project is to provide a reductive account of modality—to give an account of the truth of modal *claims* in a fundamentally non-modal world[34]—

[34] As Bricker himself says: "The problem of [the] contingency [of actuality], as I see it, is essentially a semantical problem. To solve it, the Leibnizian needs to provide plausible semantical analyses of the modal statements in question within her framework of possible worlds. Success is measured by getting the truth conditions right" (Bricker, 2006, p56).

whereas the moving spotlighter is aiming to give a metaphysic whereby *reality itself* is fundamentally tensed.

Let us look again at the counterpart theory case: just as we asked "What does a world w's being identical to itself have to do with the possibility of its being actual?," so have people asked "What does a *counterpart* of Obama's being F have to do with the possibility of *Obama* being F?" That is the essence of Kripke's Humphrey objection.[35]

The objection is that if Lewis's modal realism is true then there is not *really* any contingency in how Obama is, since he is the way he actually is at every world in which he can be found (i.e. the actual world only). However, there are two ways to understand the claim that there is not really any contingency in how Obama is: on one reading, it is simply false; on the other, it is true, but it should not be taken as an objection to Lewisian modal realism. The first reading is that it is not strictly speaking true, on Lewis's account, that Obama could have been other than he is. This is the complaint Kripke seems to have in mind when he says that on Lewis's account all that is true is that someone else could have won, not that Humphrey himself could have.[36] That is just a mistake.[37] If Lewis is right it simply *is* true—strictly and literally true—that Humphrey and Obama could have been otherwise, and the *truth-condition* for that is that they have counterparts that *are* otherwise (i.e. that are different from how Humphrey and Obama actually are). To press that version of the Humphrey objection is just to misunderstand how counterpart theory works. But there is another way of understanding the claim that there is not really any contingency in how Obama is if Lewis is right, and that is that while Lewis manages to account for the truth of the English claim "Obama could have been 4ft tall," it is not *really* the case that Obama could have been 4ft tall. What is the "really" there doing, such that it can make sense to concede that it is true that p but not *really* the case that p? The "really" here, as I will understand it, is forcing you to consider the claim as a description of how reality is at its fundamental level. So the thought behind this objection is: sure, Lewis, you have accounted for the truth of ordinary English claims concerning Obama's possibly being otherwise than he actually is, but on your view reality at its fundamental level involves no contingency in how Obama is.

That claim is correct, but it is no *objection* to Lewisian realism. The Lewisian does not *want* there to be any contingency concerning how reality is at the fundamental level. The fundamental description of reality will be a description of the *entirety* of reality: that is, for the Lewisian, a description of the space of worlds

[35] Kripke (1980, p45, fn.13). [36] Kripke (1980, p45, fn.13).
[37] As Hazen (1979) shows.

as a whole. And as Lewis explicitly admits, the true description of reality as a whole will be a non-contingent one.[38] Contingency only ever arises, on the Lewisian picture, when we are speaking parochially, describing only our mere corner of reality (what for us is actuality).[39]

For Lewis, how reality is as a whole is a non-contingent matter. The project, for him, is then to vindicate ordinary modal *judgments*: to give an account of the truth-conditions of ordinary modal claims that lets them come out true despite this underlying reality that is the way it is as a matter of non-contingency. The only criterion for success is that most of our ordinary modal judgments come out true, given how reality is according to Lewis. Lewis would not recognize as an acceptable constraint that the truth-conditions given have to look "natural," or that they respect our pre-theoretic beliefs about what it takes for our modal claims to be true. It may indeed be revisionary with respect to pre-theoretic opinion that *de re* modal claims concern what happens to mere counterparts of the *res* in other worlds, but for Lewis this is simply of no consequence: all that matters is that our ordinary *de re* modal *judgments* are not, for the most part, in need of revision given his proposed metaphysic.

Bricker's metaphysic is slightly different from Lewis's, but he is engaged in the same type of project. Bricker takes Lewis's ontology of many concrete spacetimes and adds to this something Lewis does not have: the monadic property, had by exactly one of these worlds, of *being actual*.[40] And then the project is the same as Lewis's: give an account of the truth-conditions of our ordinary modal judgments that, for the main part, has us as speaking truly when we modalize, given this metaphysic. It is surely central to our modalizing that other things could have been actual, so both Lewis and Bricker will want to secure the truth of that claim. For Lewis, this is easy, since "actual" is on his account an indexical, and every world is actual to its inhabitants. For Bricker, the task is the trickier one of saying why it comes out true that other worlds could have had this monadic property that is in fact only had by our world. And as we saw, his answer is that "w is actual" is true at v just in case w=v, so in effect "w could have been actual" is true just in case w is identical to itself. This yields the desired result that other worlds—indeed, each world—could have been actual (and that, for each world, had it been actual then no other world would have been actual also). And one can object, as we did above, that this is ad hoc, that there is no sense of *why* "w could have been actual" is a good description of modal space given the metaphysics. But

[38] Lewis (1986, p112). [39] See Cameron (2012) for discussion.

[40] There are other differences between Lewis's and Bricker's analyses of modality, but they are irrelevant to the current discussion.

really, to make this objection is to misunderstand the project: the goal is merely to have our ordinary modal judgments come out true for the most part, not to give an account of their truth-conditions that fits with any prior belief about *why* such claims are true. Likewise, one can object that Bricker's view looks like the "stuck actuality" view with a fancy semantics, not a view on which it is *really* contingent which world is actual. But to make this objection again reveals a misunderstanding of the project: Bricker's view *is* the stuck actuality view with a fancy semantics, and that is the point! As it is for Lewis, the true *fundamental* description of Bricker's reality will be a non-contingent one: contingency does not arise when you are describing reality as a whole (the space of worlds) but rather when you are describing our small corner of reality (actuality). It is simply *true* that on Bricker's view there is not *really* any contingency concerning which world is actual: meaning only that the complete fundamental account of how things are says that it is our world, @, that is actual, and this complete fundamental theory is a non-contingent one. But there is no difference here between the actuality of @ and the height of Obama. @ is actual and Obama is 6′1″: both of those are truths, and the complete correct description of reality will say that both of these things are the case; and this complete description is true and admits of no contingency, so in that sense there is not *really* any contingency in @'s being actual or Obama's being 6′1″. So there *is* a sense in which Bricker *really* has a stuck actuality view, but it is just the same as the sense in which he (like Lewis) *really* is an extreme essentialist and necessitarian: the complete account of reality admits of no contingency when it says which world is actual or what properties things have, or indeed in anything at all. But this is no objection. Or at least, if you find this objectionable then it is nothing to do with how Bricker treats actuality in particular: your objection is Williamson's objection concerning how the Lewisian-style realist treats contingency.[41] A Lewisian like Bricker is happy for the complete description of reality to be a non-contingent one: the goal is simply to reconcile this with the contingency of our ordinary modal judgments. Bricker does this: he gives an

[41] Williamson (2002, p239): "There is genuine contingency in how things are only if, once values have been assigned to all variables, the resulting proposition could still have differed in truth-value.... According to David Lewis's modal realism, contingency consists in differences between possible worlds.... Consider the common sense claim 'It is contingent that there are no talking donkeys'.... If one interprets the quantifier as unrestricted, modal realism makes the claim false... the modal realist holds that there really are talking donkeys, in spatiotemporal systems other than ours [and does not hold that this claim could have had a different truth-value]. For modal realism to make the claim true as uttered in the actual world, one must interpret the quantifier as implicitly restricted to the objects in a world.... The restricted quantifier is given an implicit argument place for a world." Williamson concludes that if Lewis's modal realism is true then there is no genuine contingency in whether there are talking donkeys—or indeed in anything at all. For discussion see Sider (2013a, pp247–8).

account of the truth-conditions of ordinary claims concerning the contingency of what is actual that has us speaking truly. That is the only criterion for success, and the fact that those truth-conditions do not obviously have anything to do with the contingency of actuality is neither here nor there.

So Bricker's realism about possible worlds with a unique objectively privileged actual world can successfully avoid the modal analog of McTaggart's argument. Bricker rejects *Possible Record* which allows him to avoid the contradiction of saying that other worlds are both actual and not actual, and he has an account of why it is true that other worlds could have been actual despite the state of affairs of their being actual being found nowhere in modal space. The account is ad hoc—but not objectionable for that, given the Lewisian project that Bricker is engaged in.

But things are different for the moving spotlighter, and she cannot avoid our version of McTaggart's paradox simply by making the Bricker move and holding that, while the state of affairs of some non-present time t's being present is not to be found anywhere in the past or future, it is nonetheless true that t was/will be present simply because the truth-conditions of that demand only the identity of t with itself. That is because the moving spotlighter does not merely want to account for the truth of our ordinary claims concerning what is present changing, but rather wants to account for the fact that *reality itself* is subject to change: that how the world is *really* used to be different, and *really* will be different.

Of course, one *could* have a project that would be satisfied with the Bricker-esque version of the moving spotlight. The B-Theorist thinks that reality is a fundamentally static place: that how things are as a whole is not subject to change. Nevertheless, the B-Theorist wishes to reconcile this static reality with the truth of ordinary claims concerning change: change is taken to be variation across this fundamentally static reality, and our ordinary claims concerning change come out true because we are ordinarily not describing reality as a whole but only our small corner of reality (the present), and we say that things have changed if things are otherwise at other small corners of reality. Suppose you share the B-Theorist's thought that reality is a fundamentally static place, and that your project is merely to account for the truth of ordinary tensed claims given this metaphysic, but unlike the standard B-Theorist you think there is a monadic objective property of *being present* that is had by one time: a property that marks this time as one out of many that is objectively privileged. On this view, which time is objectively present doesn't *really* change. But that is no surprise: on this view, *nothing* really changes, for reality is a fundamentally static place. All that matters is that our ordinary judgments concerning other things having been present, etc., come out true for the most part: and accepting the Bricker-esque proposal will secure that result.

That is the view on time that is really analogous to Bricker's view of modality. *Really*, there is a stuck spotlight: a unique, objectively privileged present that does not change. But we can nevertheless *say* that the spotlight moves: despite this fundamentally unchanging reality, we nonetheless secure the truth of ordinary claims concerning the changing of what time is present. But I take it this is simply not what most A-Theorists want out of the A-Theory; certainly, it is not what *I* want. We do not merely want to uphold the truth of ordinary tensed judgments: we want to account for the tensed nature of *reality itself*. We want the spotlight to *really* move: we want how reality is itself to be subject to change, not merely to reconcile the truth of tensed talk with a fundamentally static reality. In which case, the Bricker move is not satisfying: we want to be able to say why non-present times were or will be present, and we had better be able to say that this is a matter of how reality itself changes, and to distinguish our view from the stuck spotlight metaphysics with a semantics which allows for the truth of ordinary tensed claims.

This can be done. One option is to say that there are a bunch of brute tensed facts concerning other times having been/going to be present. We can then say that it is true that t was/will be present if and only if reality contains the brute tensed fact that t was/will be present. That will not lead us into our McTaggartian paradox, because we are not committing ourselves by saying this to saying that the presentness of t is to be found anywhere in reality. t's being present is simply a way reality was/will be, it is not a way reality *is*, even atemporally speaking. This view can easily be distinguished from the stuck spotlight with the fancy semantics, because reality contains no such brute tensed fact on the stuck spotlight view. Reality itself changes, on this view, unlike on the stuck spotlight view: an account of how reality is fundamentally will not be complete unless it mentions these brute tensed facts that describe how reality as a whole used to be and will be.

This view gives us a McTaggart-proof (since it rejects *Past Record*) theory on which reality itself is subject to change. But it is not very satisfying. It looks ad hoc: why does one kind of tensed truth, those concerned with other times having been or going to be present, get treated differently from all the others? *Past Record* (together with its future-directed counterpart, for the moving spotlighter) is true for *most* historical claims: the reason things used to be such-and-such a way is that you can find them *being* that way in the past. But when it comes to saying that all that stuff used to be present, we have a totally different account: their presentness is nowhere to be found in reality, there is merely the brute tensed fact that that way for things to be used to be present. And not only does this seem ad hoc, but in abandoning *Past Record* for one type of tensed claim, we risk undermining the whole motivation for believing in non-present ontology: if we are going to have *some* brute tensed facts, why not simply admit tensed facts for *every* truth concerning how things were or will be, in which case we might as well just be presentists. Why think that *some* historical truths must be reflected in how the past is, as *Past Record* demands, when it is not the case that *all* historical truths are so reflected?

These are serious worries, and I think that it is a constraint on any satisfying non-presentist A-Theoretic account of time that it give a satisfying answer to them. The account to be offered in chapter four aims to do just that. So to sum up, going forward the goal will be to:

1. Offer a non-presentist A-Theory that does not uphold *Past Record* (or its future counterpart) in full generality, so as to avoid the McTaggartian paradox.
2. Show how, despite not upholding *Past Record* in full generality, this theory is still more advantageous than presentism.
3. Where the theory departs from *Past Record*, show that the account given of the truth of tensed claims is not ad hoc.

To properly assess whether these desiderata are met, however, we need to be clear on what it is for there to be a brute tensed fact. We know that we must depart from *Past Record* to avoid the McTaggartian paradox: things were present, but their *being* present is not to be found anywhere in reality, even though the past is real. What departures from *Past Record* are objectionable, and what does it take to be giving an ontological underpinning for a tensed truth? Answering these questions will be one of the goals of the next chapter.

2.4 Fine's Non-Standard Realisms

Before moving on, I want to pause and look at a "McTaggart inspired" argument from Kit Fine. Fine argues that the following four theses are inconsistent:[42]

Realism: Reality is constituted (at least, in part) by tensed facts.

Neutrality: No time is privileged, the tensed facts that constitute reality are not oriented towards one time as opposed to another.

Absolutism: The constitution of reality is an absolute matter, i.e. not relative to a time or other form of temporal standpoint.

Coherence: Reality is not contradictory, it is not constituted by facts with incompatible content.

[42] Fine (2005, p271).

The idea behind *Realism* is this. Everyone agrees that some tensed claims are true. It is true (at the time of writing, at least) that Obama is now president and that Bill Clinton was president but is not so now. But what is up for debate is whether these tensed truths are sensitive to a reality that is ultimately tensed. Fine's position is that in order to ask such questions we need a primitive notion "In reality..." Something can be the case without being the case *in reality*. 2+2 is, undoubtedly, 4. That mathematical truth is something that is the case. But there is a further question as to whether in reality 2+2 is 4. Is reality constituted by distinctively mathematical facts, or is the truth of mathematical claims sensitive to a realm of non-mathematical facts (such as, for example, modal facts concerning the possible existence of certain structures)? The realist about tense thinks that reality is constituted by facts like Obama's now being president. The *now*-ness is involved in the facts themselves, and so reality itself is tensed. The anti-realist, by contrast, says that the truth "Obama is now president" is sensitive not to a tensed fact but to the untensed fact of Obama's being president from 2008 to 2016. The facts that constitute reality, according to the anti-realist, are atemporal ones, involving no *now*-ness or *past*-ness, etc.

Fine argues that the four principles just outlined form an inconsistent tetrad, and hence the realist about tense is forced to deny one of the latter three principles, with a different style of realism resulting, depending on which principle is denied. Here is my reconstruction of the argument for the inconsistency.

The following two claims are undeniably true. Bill Clinton is now not president. Bill Clinton was president. From the first claim, *Realism* tells us that reality is constituted in part by the tensed fact that Clinton is now not president. From the second claim, *Realism* tells us that reality used to be constituted by the tensed fact that Bill Clinton is now president. (It maybe also tells us that reality is now constituted by the tensed fact that Bill Clinton was president, but that is not what is going to drive the argument for the inconsistency: we are focusing on what this historical truth tells us about how reality used to be constituted, not whether it tells us that reality is now constituted by facts about how things used to be.)

Neutrality tells us that no time is privileged. That is to say, there is no privileged temporal standpoint from which you can give the "correct" description of what facts constitute reality. That means that if reality *was* constituted by certain facts, those facts must be involved in how reality *is* constituted: at the very least, we must admit that reality *is* constituted by those facts *at that past time*. To say otherwise would be to ignore the reality of the way things were at that time. *Realism* made us conclude that reality *used* to be constituted by the fact that Clinton is now president. *Neutrality* is now going to force us to conclude that this fact is involved in how reality *is* constituted: at the very least, reality is constituted *at a past time* by the fact that Clinton is now president.

Absolutism tells us that there is such a thing as the facts that constitute reality simpliciter. If asked what facts constitute reality, I should be able to answer simply by giving a list of them. I do not need to qualify my answer by saying, for example, well these are the facts that constitute reality at *this* time, and these are the facts that constitute reality at *that* time, etc. I can simply say which facts constitute reality—end of story. *Absolutism* tells us that if reality is constituted *at a time* by a certain fact, then it is constituted simpliciter by that fact. The "at a time" is doing no work, because reality is not constituted *at a time*, it is constituted simpliciter. Compare the spatial case. Reality is constituted in Edinburgh by the existence of the Scottish parliament. But reality is not constituted relative to places: how reality is *somewhere* is simply part and parcel or how it is simpliciter. So reality is constituted, simpliciter, by the existence of the Scottish parliament. Likewise, if *Absolutism* is true, reality is not constituted relative to times: how reality is *at a time* is simply part and parcel of how it is simpliciter. So since *Neutrality* told us that reality is constituted at a past time by the fact that Clinton is now president, *Absolutism* now tells us that reality is constituted, simpliciter, by the fact that Clinton is now president.

But right at the start, *Realism* told us that reality is constituted by the fact that Clinton is now not president. So reality is constituted, simpliciter, both by the fact that Clinton is now president and by the fact that Clinton is now not president. But these facts are incompatible: these are incompatible ways for Clinton to be. So reality is constituted by incompatible facts. But this conflicts with *Coherence*, which demands that the facts that constitute reality not have incompatible content. And so we have an inconsistent tetrad, and so the realist about tense must deny one of *Neutrality, Absolutism*, or *Coherence*.

To deny *Neutrality* gives us what Fine calls standard realism, and which as I have been using the term amounts to the A-Theory. The standard realist, or A-Theorist, thinks that there is a privileged time—the present—and hence a privileged temporal perspective—that is, the present perspective—from which one can give the correct account of the facts that constitute reality. According to the standard realist, the fact that Clinton is now president is simply not amongst the facts that constitute reality. That fact *was* amongst the facts that constituted reality, but that is simply how reality *used* to be constituted, it is not how reality is constituted simpliciter. How reality is constituted, simpliciter, is how reality is constituted *now*, says the standard realist. The facts that used to constitute reality simply do not get in on the act.

Fine prefers, however, for reasons we will come to, the non-standard versions of realism: those theories you get by maintaining *Neutrality* but denying either *Absolutism* or *Coherence*. As I am using the term, these theories do not count as A-Theories, since they deny that there is a privileged present. Thus the A-Theory and the B-Theory are not exhaustive options: one can be a realist about tense (and hence not a B-Theorist), without believing in a privileged present (and hence not being an A-Theorist). Fine has done us a great service by making these alternative options clear.

To deny *Absolutism* is to be a relativist. According to this form of relativism, there is no such thing as *the facts that constitute reality*. Reality is not constituted simpliciter, it is only constituted relative to a time, and so there are only the facts that constitute reality relative to *this* time, the facts that constitute reality relative to *that* time, etc. The relativist will grant that the fact that Clinton is now president constitutes reality relative to 1995 and does not constitute reality relative to 2015, and that the fact that Clinton is now not president constitutes reality relative to 2015 and does not constitute reality relative to 1995. There is no time relative to which reality is constituted by both the fact that Clinton is now president and the fact that Clinton is now not president. And when asked whether reality is constituted by the fact that Clinton is now president, the relativist will respond that this question is not well formed for it is missing a parameter: reality is only constituted or not by a fact relative to a time. Since there is no time relative to which reality is constituted by facts with incompatible content, the relativist holds on to the coherence of reality.

This relativist view is very different from standard realism. Both the relativist and standard realist will say that reality is now constituted by the fact that Clinton is now not president and that reality *was* constituted by the fact that Clinton is now president. But they have a very different understanding of what this amounts to. The standard realist, in making a claim about how reality is *now* constituted, is making a claim about how reality is constituted *simpliciter*, and in making a claim about how reality used to be constituted she is making a claim about how it used to be constituted simpliciter, but she takes this to have no consequences on how reality is in fact constituted. By contrast, the relativist thinks there are simply two equally correct accounts of reality's constitution: relative to one time a certain set of facts, including that Clinton is now president, constitute reality, and relative to another time a certain set of facts, including that Clinton is now not president, constitute reality. Neither set are the "correct" set of facts that constitute reality for there is no such thing as the facts that constitute reality simpliciter: each set constitutes reality relative to the relevant time, and that is all that can be said about how reality is constituted.

It is important also to contrast Finean relativism with certain comparatively innocuous views that go by the name "relativism." Some people are attracted to relativism about attributions of, for example, personal taste, so that the sentence "Rhubarb is tasty" can be true for me and false for you, for we each evaluate it for truth or falsity relative to our different standards of taste.[43] But this is *not* (or at least, it need not be; and this is not what defenders of this relativism are committing to) a view on which reality is constituted by facts about taste, but where there is no such thing as the facts that constitute reality *simpliciter*, merely the facts that constitute reality relative to me and the facts that constitute reality relative to you. That metaphysic would be hard to swallow indeed.[44] The more plausible account of what is going on if attributions of taste are relative in this sense is that reality is not constituted by facts about taste at all; rather, claims about tastiness are sensitive to other facts (such as facts about the chemical makeup of foods, facts about the effects of such foods on people with such-and-such taste buds, etc.), and the best *semantic* account of how sentences that make attributions of taste are sensitive to these facts is one which takes the truth of those sentences to be relative to a standard of assessment. This relativism, then, is a semantic doctrine, not a metaphysical one: it is the *truth-value* of *sentences* that is relative, not the underlying reality that the truth of those sentences is sensitive to. The relativist about taste can give an answer to the question of how reality is *simpliciter*—she just won't use claims about tastiness in doing so. Fine's tense relativism is much more radical: it is reality itself that is relative, not the truth-value of certain representations of reality. *How reality is*—what the underlying facts are that constitute reality and that *make* representations of reality true or false—is relative to a time.[45]

To deny *Coherence* is to accept that reality is constituted, simpliciter, by both the fact that Clinton is now president and the fact that Clinton is now not president. That is to embrace the claim that reality is ultimately an incoherent place, being constituted by facts whose content is incompatible. Fine calls this view "Fragmentalism," because while reality can be parceled out into coherent *fragments*, reality as a whole is incoherent.

[43] See *inter alia* Egan (2010).

[44] Unlike rhubarb, which is easy to swallow because it is in fact delicious.

[45] MacFarlane's relativism about time (2003, 2008) is most naturally read as a semantic doctrine, not akin to the radical metaphysical view that Fine is defending. MacFarlane is defending the comparatively innocuous view that the truth-value of tensed *claims* is relative to a feature of our context of assessment, namely, what time we are making the assessment at. This is compatible with the B-Theoretic claim that reality consists of an underlying domain of absolute tenseless facts.

It is tempting to see Fragmentalism as a species of dialetheism, the view that there are true contradictions, and to see the Fragmentalist thereby as holding both that Clinton is now president and that he is now not president. But that would be a mistake. In some ways, it is the opposite of dialetheism: dialetheism is compatible with reality being consistent; it is solely a claim about the inconsistency of *truth*, while fragmentalism is compatible with every truth being consistent; it is solely a claim about *reality*. Dialetheism is a semantic thesis: there are true representations of reality that also have true negations.[46] Fragmentalism is a metaphysical thesis: the facts that constitute reality, to which the truth-values of representations are sensitive, contradict one another. These theses go hand in hand if we assume that there can be true contradictions if and only if there are incompatible facts constituting reality to make that contradiction true. But that assumption is not forced upon us: a dialetheist can deny it, and Fine (when defending Fragmentalism) does deny it. The biconditional can be denied in each direction: we might accept that reality is consistent but that it nonetheless gives rise to true contradictions, and we might accept that an inconsistent reality can nonetheless never truly be described by a contradictory sentence.

Of course, there is a *sense* in which reality must be inconsistent if there is a true contradiction, insofar as reality must be such as to make that contradiction true.[47] But if something stronger is meant by "reality is inconsistent" (if, for example, we are happy to follow Fine in distinguishing what is the case from what is the case *in reality*[48]) then accepting that some contradiction truly describes reality does not commit us to accepting that reality is inconsistent (in Finean terms, a contradiction can be the case without any contradiction being the case in reality).[49] One can accept the semantic claim that there are true contradictions

[46] Mares (2004) distinguishes between *metaphysical* and *semantic* dialetheism. I am making use here of the same distinction, although I prefer to reserve the term "dialetheism" purely for the semantic thesis. Partly that is because the reference to truth—a semantic notion—is built into the very name; but also because Graham Priest surely gets a large say in dictating how the term is used, and he uses it solely to refer to the semantic thesis: "Dialetheism is simply the view that some contradictions are true. That is, there are some sentences . . . such that both [they are their negations] are true. This is a view *about language*" (Priest, 2006, p299, my emphasis).

[47] As Priest (2006, p299) admits.

[48] Fine (2000).

[49] As Priest (2006, p302) says, whether or not you ought to conclude from the truth of a contradiction that *reality* is inconsistent in some metaphysically substantive sense will likely depend on the extent to which you are a metaphysical realist about the subject matter of the contradiction. If you think that any complete and true account of what is legal, e.g., will be contradictory, you may well resist the conclusion that *reality* is inconsistent, on the grounds that the legal realm is in some sense a human construct. As Priest puts it "legal dialetheias [are] the result of simple say-so. There is no language-independent reality in the appropriate sense" (Priest 2006, p302). By contrast if you think you cannot truly and completely describe what it is for a thing to be in motion without

without accepting the metaphysical claim that the underlying facts that make true what is true are inconsistent. It is a perfectly intelligible thesis that reality itself is consistent but that this consistent reality leads to an inconsistent assignment of truth-values to representations of reality.

Likewise, believing that reality is an inconsistent place does not by itself entail that there are true but contradictory representations of reality. Fine's Fragmentalism denies this entailment, and thus the view is not a species of dialetheism. Here is what Fine says:

> Although there is a sense in which the fragmentalist takes reality to be contradictory, her position should not be seen as an invitation to accept contradictions. Even if reality contains both the fact that I am sitting and the fact that I am standing, it will not be correct for me simultaneously to assert both that I am sitting and that I am standing. For any such assertion will only relate to those aspects of reality that "cohere" with the existence of the given assertion; and so, it will only be correct for me to assert that I am sitting if, at the time of the assertion, I am sitting.[50]

As I understand it, what is going on is that reality is constituted by inconsistent facts, but we deny the version of the correspondence theory of truth that says that a claim is true if it corresponds to some facts in reality. Reality might be constituted by facts that correspond to some representation of reality without that representation being true. That is because representations of reality, and utterances that express such representations, are to be found in consistent *fragments* of reality, and whether or not they are true or false depends on whether there is a corresponding fact that constitutes *that* fragment of reality. Since each fragment is consistent, no utterance of a contradictory sentence—no inconsistent representation of reality—will ever be true; hence, dialetheism will be false. In the present case, reality is constituted both by the fact that Clinton is now president and by the fact that Clinton is now not president. But whether a representation of reality that represents Clinton being now president is true depends not on whether reality is constituted partly by the fact that Clinton is now president but rather whether the fragment of reality that contains the representation is partly constituted by that fact. Since no fragment of reality is constituted by both the fact that Clinton is now president and the fact that Clinton is now not president, no representation that represents Clinton being both now president and now not president will be a true one, and no utterance of the sentence "Clinton is now president and now not president" will be true. So we have the

contradicting yourself (Priest, 2006, ch.12)—well, things move no matter what *we* have to say about it, so that looks like a *prima facie* case of reality in and of itself being inconsistent.

50 Fine (2005, p282).

metaphysical inconsistency without the semantic inconsistency: an inconsistent reality that can never truly be described by an inconsistent sentence.[51]

So those are Fine's two versions of non-standard realism about tense. I am a committed standard realist: I avoid Fine's inconsistent tetrad by denying *Neutrality*. There is a privileged present, the time that is now, and the facts that constitute reality simpliciter are the facts that constitute it *now*.[52] I find the metaphysics of both relativism and Fragmentalism very counter-intuitive. I am strongly inclined to the view that there is a way reality is, simpliciter, and that it is a consistent way. But Fine offers three arguments that you should be a non-standard realist, of either variety, if you are going to be a realist about tense at all. One concerns the non-standard realist being better able to reconcile her view with the findings of special relativity. I will not address this argument: as I said in the introduction, I leave the battle of reconciling the A-Theory with current physics to others. But let me say something about Fine's other two arguments for preferring non-standard realism.

The first argument concerns that most elusive of notions, temporal passage. Fine argues that standard realism cannot account for the fact that time *passes*, but that this is so in virtue of features of that view that are not shared by the non-standard realisms. Thus, while he does not actually aim to show that either version of non-standard realism *does* account for passage, he thinks he has at least shown that they do not face the same *barrier* as standard realism does in accounting for passage. As such, there is an objection to standard realism that does not threaten non-standard realism, and this is a reason to prefer the latter. Here is Fine:

> The standard realist faces a general difficulty. For suppose we ask: given a complete tenseless description of reality, then what does he need to add to the description to render it complete by his own lights? The answer is that he need add nothing beyond the fact that a given time t is present, since everything else of tense-theoretic interest will follow from this fact and the tenseless facts. But then how could this solitary "dynamic" fact, in addition to the static facts that the anti-realist [i.e. the B-Theorist] is willing to accept, be sufficient to account for the passage of time?... [The standard realist's] conception of temporal reality, once it is seen for what it is, is as static or block-like as the anti-realist's, the only difference lying in the fact that his block has a privileged centre. Even if

[51] This does raise the worry that Fragmentalism cannot be coherently stated. If an utterance is true iff it corresponds to the facts that constitute the consistent fragment of reality that contain that utterance, then seemingly no utterance of "Reality consists of multiple consistent fragments that are jointly inconsistent" will be true. The threat is that if Fragmentalism is true, it cannot be truly said to be so. The Fragmentalist may have to associate with Wittgensteinians (to the former's discredit).

[52] If you are worried about how this is compatible with being an eternalist, I hope to answer those concerns in chapter four.

presentness is allowed to shed its light upon the world, there is nothing in his metaphysics to prevent that light being "frozen" on a particular moment of time.[53]

The objection, as I understand it, is this: one gets a standard realist (A-Theorist) metaphysic by taking a B-Theoretic tenseless metaphysic and adding to it the feature that a particular time is objectively present. From this, all the other features of tensed reality—such as what times were or will be present—are meant to follow. But where is the *passage*? All we have done is add a property to a point of a static metaphysic—how can that have introduced something *dynamic*? Part of Fine's objection seems to be that the standard realist has no right to say that it is an objective feature of reality that an earlier or later time *was* or *will be* present; that this will just be a way of speaking. He says:

> The [standard] realist at this point might appeal to the fact that any particular future time t+ *will be* present and that any particular past time t− *was* present. However, the future presentness of t+ amounts to no more than t being present and t+ being later than t and, similarly, the past presentness of t− amounts to no more than t being present and t− being earlier than t. But then how can the passage of time be seen to rest on the fact that a given time is present and that various times are either earlier or later than that time?[54]

Each version of non-standard realism is meant to avoid this problem because they are *not* simply taking a B-Theoretic metaphysic and privileging a point in it: each view has presentness applying in some sense to *every* time, so *all* times are objectively privileged in some way. And thus—I take it the thought is—there is no barrier to saying that presentness objectively applies to each time *successively*. Fine says:

> The two forms of non-standard realism are not subject to these difficulties since they do not single out any one time as *the* present. For the external relativist, each time is objectively present at that time: at each time t, reality is constituted by the absolute fact that t is present.... And for the fragmentalist, each time t is objectively present *simpliciter*... [I]n either case, presentness, in so far as it is a genuine feature of reality, applies equally to all times. Presentness is not frozen on a particular moment of time and the light it sheds is spread equitably throughout all time.[55]

I think that Fine's objection to standard realism here basically amounts to a challenge to the A-Theorist that I conceded earlier: that she had better be able to distinguish her view from the stuck spotlight metaphysic coupled with a fancy semantics that lets her *say* that other times were or will be present. (Cf. Fine's complaint that "there is nothing in [the standard realist's] metaphysics to prevent

[53] Fine (2005, p287). [54] Fine (2005, p287). [55] Fine (2005, pp287–8).

that light [presentness] being 'frozen' on a particular moment of time."[56]) Both non-standard realisms recognize the reality of some past (or future) time t's being present: for the relativist that amounts to reality being constituted *at that time* by the fact that t is present; for the fragmentalist, reality is constituted simpliciter by the fact that t is present. But the standard realist cannot, on pain of contradiction, admit the reality of t's presentness. They cannot say that t is present *in the past* [/future], because that leads to McTaggart's paradox. They have to simply hold that reality in no way includes the presentness of t. Of course, they will say that reality *did* [/will] include the presentness of t. But the challenge is: why is this not just talk? Why are they not simply allowing themselves to speak a certain way: to *say* "t was present" iff t is before a time that *is* present? One can stipulate that those are the truth-conditions of "t was present" if one wishes, but allowing yourself to speak a certain way does not affect the *metaphysics*, and there is nothing in the picture so far to distinguish this view from the stuck spotlight view.

The moral, as I see it, is that the standard realist, or A-Theorist, should *not* simply think of their metaphysic as taking the B-Theorist's metaphysic and simply adding to it that some point in time be objectively present, with all other tensed facts to follow from this. To do that *would* be to simply have a stuck spotlight metaphysic. The A-Theorist needs to be able to say more about what makes it the case that some past time *was* present than that it is before some time that *is* present, but (to avoid McTaggart's paradox) she needs to do this without saying that that past time *is* present in the past. She needs to steer a middle ground: the past presentness of a time needs to be a genuine feature of reality and not just a way of speaking, to distinguish the view from the stuck spotlight view, but this time's being present cannot be a way the past *is*, so as to avoid McTaggart's paradox. The metaphysic to be presented in chapter four aims to steer exactly this middle ground.

Fine's second argument for non-standard realism over standard realism builds on a familiar argument from Gareth Evans[57] concerning the relationship between reality and the contents of our thoughts and assertions, so let us begin by looking at that argument. The A-Theorist is committed to the view that at least some of our thoughts and utterances have contents whose truth-value changes. To think otherwise would be to hold that the distinctive tensed features of reality are simply inexpressible in thought or language, which would put the A-Theorist in the embarrassing position of not being able to contemplate or state her own theory. So if the A-Theory is true, there are some things we think and say that are true but will be false, and so some of our thoughts and utterances are correct but

[56] Fine (2005, p287). [57] Evans (2002).

will be incorrect. But Evans says that this is "such a strange position that it is difficult to believe anyone has ever held it."[58] Why? Because

> [W]e use the term 'correct' to make a once-and-for-all assessment of speech acts.... [I]f a theory of reference permits a subject to deduce merely that a particular utterance is now correct but later will be incorrect, it cannot assist the subject in deciding what to say, nor in interpreting the remarks of others. What should he aim at?... Maximum correctness?[59]

Here is Jeffrey King making a similar claim:

> [T]hough it seems correct to hold that the things I believe, doubt, etc. can change truth value across worlds (i.e. some of the things I believe are true though they would have been false had the world been different), it is hard to make sense of the idea that the things I believe may change truth value across time and location. What would it be e.g. to believe that the sun is shining, where what I believe is something that varies in truth value across times and locations in the actual world? It seems clear that when I believe that the sun is shining, I believe something about a particular time and location, so that what I believe precisely does not vary in truth value over times and locations. Right now I am in Santa Monica looking out the window. I believe the sun is shining. Is it really credible to think that my belief is not about Santa Monica now?[60]

One initial comment: it is perfectly consistent with the A-Theory that ordinary thoughts and utterances do *not* have contents whose truth-value is subject to change.[61] Even given an A-Theoretic metaphysic, it might be the case that when I have an ordinary belief like the belief that the sun is shining, what I believe is the tenseless claim that the sun is shining *at* the particular time at which I am having the belief. The A-Theorist might be right about how the world is but the B-Theorist right about the contents of our ordinary thoughts and assertions. At most, what the A-Theorist is *committed* to is that we *can* think and assert things with contents whose truth-value is subject to change; but it may be the case that such contents are only thought or asserted in extraordinary conditions, such as when we are in the metaphysics classroom discussing the tensed nature of reality. What seems clear to King, then, might well be the case: when he believes that the sun is shining, he believes something about a particular time t: a tenseless content of the form that the sun is shining *at t*. That in itself does not mean that there is not another thought to be had: that the sun be shining at *whichever* time is

[58] Evans (2002, p349). [59] Evans (2002, p349).

[60] King (2007, p166). See also King (2003). I will ignore King's mention of locations in what follows. I *do not* think there are thoughts whose truth-value varies across spatial location, even though I think there are thoughts whose truth-value varies across times—that is because I think there is a metaphysical asymmetry between time and space: how reality is as a whole changes from moment to moment in time, but it does not change from point to point in space.

[61] Though see Sullivan (2014) for an account that resists even this mild conclusion.

objectively present (a content whose truth-value changes as what time is present changes).[62]

Now consider Evans's complaint that if we allow contents to have temporary truth-values our theory "cannot assist the subject in deciding what to say, nor in interpreting the remarks of others."[63] That seems too strong: if what we say can change in truth-value then, seemingly, a subject should aim to say things that are true when they are said, and she should interpret others as saying things they believed to be true when they said them. What should one *aim* at with one's assertions, asks Evans: why is not "truth at the time of utterance" a perfectly good answer if one's assertions have truth-values that are subject to change? Of course this might be a bad theory of how communication in English works: it might be a mistake to interpret ordinary speakers in ordinary contexts as committing merely to their utterances being true when they are uttered, and "truth at the time of utterance" might be a bad hypothesis about the aim of the utterances of ordinary speakers in ordinary contexts. But that speaks only to the issue of whether ordinary thoughts and utterances have temporary truth-values and, as I said above, the A-Theorist simply does not need to fight that battle. Let the B-Theorist have her account of how ordinary language works; that is simply tangential to the metaphysical issue.

Of course, Evans and King seem to be suggesting something stronger: that there is something *incoherent* about the notion of a content whose truth-value is subject to change. King says this idea is "hard to make sense of," and asks "what would it be" to believe such a content?[64] And Evans says that we use "correct" to make a once-and-for-all assessment of speech acts, thus ruling out that

[62] That said, I would resist the implication King makes that if he is indeed having the thought with a temporary content that his belief is not thereby "about Santa Monica now." There are two ways your belief can be about Santa Monica now: because it is now time t, and your belief has a tenseless content of the form "Santa Monica is F at t," or because it is now time t, and your belief has a tensed content of the form "Santa Monica is F." The content you believe if you believe the latter will not remain about that time, of course—as time passes, that very content will be about Santa Monica at different times. But nonetheless, one will always be able to truly say that that thought is about Santa Monica now (since it is about Santa Monica at *whatever* time happens to be now). So what King thinks is not credible is, I think, no implication of our thinking the A-Theoretic content.

[63] Evans (2002, p349).

[64] King (2003, p196), also (2007, p166). Interestingly, King goes on to allow that we *can* think temporary contents. He says "I think we often talk about temporally neutral contents, as when you say 'America is the moral leader of the world' and I respond 'That might have been true ten years ago, but it isn't now.' I might have even responded 'I believed that ten years ago but I don't today.' But despite this I think that the objects of our attitudes and what we assert etc. are not temporally neutral" (King, 2007, p188, fn.36). Here it sounds very much as though King agrees with my claim that while temporary contents might not be the content of our ordinary thoughts and utterances, they are there to be thought at least. So I suspect his talk of this being "hard to make sense of," and his asking "what would it be" to believe such things is actually stronger than anything he wants to

something we say be correct at one time but false at another, as would have to be the case were its truth-value subject to change. I have to confess, I simply do not understand this complaint. I have tried to think myself into the headspace from which it seems worrying, and I have failed. Evans's claim simply seems to beg the question: if I say something whose truth-value is subject to change, then when I say it is correct I am simply *not* making a once-and-for-all assessment. If a fellow A-Theorist tells me that lunar colonies are mere future entities, I will say they are correct, and both of us will understand that this assessment will not remain true once the time at which lunar colonies are built becomes present. Evans might be right—I have no opinion on this—in thinking that in ordinary everyday speech we use "correct" to make a once-and-for-all assessment of speech acts; as I said, the B-Theorist might be right about how to treat ordinary everyday assertions. But if he is making the stronger claim that it is *incoherent* to judge utterances as temporarily correct, then that just begs the question. Similarly, when King asks "what would it be" to believe something that is true (say) but will be false—well, just that! I do not understand why this is "hard to make sense of." It is to believe that reality is a certain way, but that it will not always be that way. To find incoherence in that is simply to find incoherence in the idea of reality as a whole being subject to change—that is, to find incoherence in the A-Theoretical metaphysic itself.

Let us now look at Fine. Fine sets up his Evans-esque puzzle with the following inconsistent triad:[65]

> *Link*: An utterance is true if and only if what it states is verified by the facts (in reality).
>
> *Truth-Value Stability*: If an utterance is true (false), then it is always true (false).
>
> *Content Stability*: If an utterance states that P, then it always states that P.

These three principles generate an inconsistency as follows. Suppose I am now standing but was sitting. At both times I uttered the sentence "I am standing." Each utterance, we will assume for the sake of argument,[66] expresses the same content: the *tensed* proposition that I am standing—that is, a proposition that says nothing about the particular time at which the utterance expressing it was

commit to. But in the main text I will take him at his word and interpret him as thinking there is something dubious about the very notion of tensed contents.

[65] Fine (2005, pp289–90).

[66] Fine goes on to offer a version of the argument that does not make this assumption, but I will not go through it here. I am happy to grant him this assumption, and what I say in response to this version of the argument would apply equally to the other one.

made. By *Link* the utterance I make now is true. *Ex hypothesi*, I am now standing, so reality is constituted by the fact that I am standing, so my utterance of the sentence "I am standing" states something that is verified by the facts, and hence is true. So by *Truth-Value Stability*, that utterance is always true; *a fortiori* that utterance was true at the time at which I was sitting. By *Content Stability* that utterance stated then what it states now, namely, that I am standing. Since my earlier utterance also states this then it must also be true at that time; by *Link*, the truth of utterances is sensitive only to whether what it states is verified by the facts, so two utterances cannot state the same thing and have different truth-values. So my earlier utterance of "I am standing"—the one made when I was sitting—was true at the time I made it. But that utterance was *not* true: one does not truly utter "I am standing" if one says it when one is sitting. So that utterance is both true and false: contradiction.

It should be clear from what I said above that I wish to avoid this contradiction by rejecting *Truth-Value Stability*. I think that utterances of tensed propositions change their truth-value as how reality is changes. But Fine anticipates this move and says that it does not matter if you think there is a notion of truth that is *not* stable; all that matters is that there is some notion of truth that *is* stable, and he thinks that there is. So long as there is *a* notion of truth that obeys *Truth-Value Stability*, we can run the above argument using *that* notion and arrive at a contradiction.

I confess, I am not sure what Fine means by "a notion of truth" here. Is the argument against standard realism assuming pluralism about truth—the kind of view defended by Crispin Wright, *inter alia*, that there is more than one property of being true, and hence that different propositions can be true in different senses of "true"?[67] If so, that substantially weakens the argument; it was not presented as relying on any controversial claims concerning the metaphysics of truth. I am happy to grant that there are some *claims* (some propositions, thoughts, or utterances) whose truth-value is stable: but it is not because there is a distinctive, stable, notion of *truth* that applies to these claims, but rather it is the same notion of truth that sometimes but not always applies to other claims, but stably applies to these claims simply because they concern matters of fact that always obtain. Assuming that the laws of nature are not variable, for example, a statement that such-and-such a law obtains will have a stable truth-value. But it is the very same notion of truth that applies to the statement of the law that applies to the statement that it is raining: there is one notion of truth that applies stably to the former and non-stably to the latter simply because of the difference in subject matter—whether it is raining changes, but whether a law obtains does not.

[67] See *inter alia* Wright (1992).

Fine does not make explicit any assumption of pluralism about truth, so perhaps this is misinterpreting him. Let us look at what he says, to see what we can make of it. He offers two considerations in favor of there being "a stable notion of truth." First he says:

Our ordinary notion of truth, as applied to utterances, appears to be stable. Suppose I utter the words "I am sitting" while sitting; and suppose that a few minutes later I stand up. Someone may then ask "is that utterance KF made five minutes ago true?". The correct answer is surely "Yes", despite the fact that I am now standing.[68]

My response to this is the same as my response to Evans. Perhaps Fine is right that the answer to this question is "Yes." If so, I say, that simply means the B-Theorist is right about the contents of our ordinary assertions. For if Fine actually expressed the tensed proposition that he is sitting with his utterance of "I am sitting," as opposed to the untensed proposition that he is sitting at the time of utterance, then the answer would simply be "No—the utterance was true when he made it, but that very utterance is now false." As I said, I have nothing at stake in this debate concerning how ordinary language works, so long as the tensed propositions are there to be expressed. Now Fine goes on to say that "We cannot sensibly say that [the utterance] was once true or that it is no longer true."[69] If this is meant to mean that it is *incoherent* to suppose that someone uttered a tensed proposition, and hence that their utterance was true but is now false, then I respond as I did to King above: I see no reason at all to suppose that this is the case. (And Fine has no more to say on this point; he simply proclaims this to be something you cannot sensibly say.)

The other consideration Fine offers is meant to bypass any concerns about our ordinary notion of truth. Here is the entirety of what he says:

It will be agreed that the truth of a *current* tensed utterance requires that reality be tensed. Now there is a sense of "true" (whether the ordinary sense or no) in which it will still be correct to say a moment later in time that the utterance is true. But surely we are unable to discern any metaphysical difference between the truth of the utterance at the one time and at the other time. In so far as we are inclined to say now that the truth of a current utterance requires that reality be tensed, then surely we are equally inclined to say a moment later that its truth requires that reality be tensed. We have no sense of the metaphysical ground for the truth of the utterance shifting under our feet, as it were, as we go from the one time to the other.

What these considerations bring out is the way in which we are willing to adopt an eternal perspective of what the truth of a tensed utterance might require of reality. The requirement is the same whether we consider the truth of the utterance at one time or at another;

[68] Fine (2005, p293). [69] Fine (2005, p294).

and to the extent that this is so, it is impossible for the realist to evade the argument by appeal to the instability of truth.[70]

I find the argument in these passages very elusive. Let us focus on a particular current tensed utterance: my utterance now of "It is raining." And let us assume that this expresses the tensed proposition that it is raining in the privileged present, not the untensed proposition that it is raining at the particular time at which I made the utterance. I do indeed agree that the truth of this requires that reality be tensed. But Fine's very next claim, that "there is a sense of 'true' (whether the ordinary sense or no) in which it will still be correct to say a moment later in time that the utterance is true"[71] is only correct, I think, if it is still going to be raining in a moment's time! If the weather suddenly changes then I simply do not admit that there is *any* sense of "true" in which this tensed utterance will still be true (or that it will still be correct to say that it is true). After all, that utterance says that it is raining, but it will not be; so what it says will not be the case; so it will not be true. So I see no reason to agree with Fine's claim here. I agree with what he *goes on* to say—but I do not see how it is meant to speak to that claim he just made. Fine says: "In so far as we are inclined to say now that the truth of a current utterance requires that reality be tensed, then surely we are equally inclined to say a moment later that its truth requires that reality be tensed."[72] I agree. The truth of "It is raining" always requires that reality be tensed—but this does not mean that it is always true if it ever is, of course; for as well as always requiring that reality be tensed, its truth also always requires that it be raining. But it is not always raining, so it is not always true. Likewise, I agree with Fine that "we are willing to adopt an eternal perspective of what the truth of a tensed utterance might require of reality. The requirement is the same whether we consider the truth of the utterance at one time or at another."[73] Sure: the *requirements* that the truth of an utterance demands of reality remain the same from one time to another. The reason the truth-value of the utterance changes is not because what its truth *requires* of reality changes, but rather because whether reality *meets* those requirements changes. The truth of "It is raining" always requires of reality the same thing: that it be raining. But whether reality *meets* those requirements—and hence whether the utterance is true—changes, because whether or not it is raining changes. So I simply do not see why Fine concludes from this that "it is impossible for the realist to evade the argument by appeal to the instability of truth."[74] She evades the argument by appealing to the instability

[70] Fine (2005, p294). [71] Fine (2005, p294). [72] Fine (2005, p294).
[73] Fine (2005, p294). [74] Fine (2005, p294).

of truth because she holds that while the truth of an utterance makes a stable *demand* on reality, what is unstable is whether reality meets those demands—and hence, the truth-value of the utterance itself is unstable.

To conclude, I see no reason at all to abandon standard realism for these Evans-esque reasons. Perhaps our everyday utterances have stable truth-values, and perhaps not, but that is neither here nor there as concerns the metaphysics. The truth of tensed propositions is not stable, and I have seen no reason to think that this is problematic. As to passage, there is indeed a challenge for the standard realist—to distinguish their view from the stuck spotlight view—and to this challenge we shall return in chapter four.

3

On Giving an Ontological Account of Tense

One goal of the metaphysic to be defended in this book is to give an ontological account of tensed truths. But what does that mean? In this chapter I consider two alternatives. There is what we might call the *ideological project*: to say *what it is* for each tensed truth to obtain. Or there is the *truthmaking project*: to say what *makes true* each tensed truth. The central goal of this chapter is to show that many of the problems leveled against the moving spotlight result from pursuing the ideological project, and to thereby motivate the alternative truthmaking project. In §3.1 I consider the ideological project as viewed through the lens of the meta-metaphysics developed in particular by Theodore Sider. I agree with Sider that viewed thus, the moving spotlight does not look particularly attractive. In §3.2 I defend the alternative truthmaking project, that presupposes a meta-metaphysics closer to the theory of grounding defended by Kit Fine. In §3.3 I argue that this gives us an alternative way of thinking about what it takes to give an ontological account of tensed truths, drawing analogies to cases of modality, predication, and parthood. We will enter chapter four with the goal of saying not what it is for a tensed claim to be true, but merely what in the world makes the tensed truths true. From where in our metaphysic does tense arise?

3.1 The Quine–Lewis–Sider Picture

We ended chapter two with a challenge to any non-presentist A-Theory. In order to avoid the McTaggart-esque paradox, *Past Record* (If something was the case, then it is the case in the past) must be denied in its full generality. In particular, that some time or event was or will be present cannot be said to obtain in virtue of that time or event's *being* present at some time in the past or future. There must be another explanation for the fact that things were or will be this way other than that the extant past or future is a witness to things being that way. This raises the specter of two problems. First, there is the worry that whatever we say will be ad hoc—that we will be treating truths of the form "A was/will be F" one way when F is anything other than an A-property, but when it comes to saying that something had or will have a different A-property from that which it now has we invoke a completely different account; for example, we invoke *Past Record* for most tensed claims, but postulate brute tensed facts to ground the change in A-properties. Second, there is the worry that we will undercut (at least some of) the motivation for believing in non-present ontology in the first place. Part of the job mere past and future entities were supposed to be doing was providing an ontological underpinning for tensed truths, but if we have to look elsewhere to provide an account of the different A-properties things used to have, could we not have simply used whatever resources we appeal to for that to provide an account of all tensed truth, thus rendering non-present ontology redundant (in this respect at least)? If we must invoke brute tensed facts, for example to ground changes in A-properties, can we not simply be presentists and invoke them to ground all the facts about how things were?

To know whether a metaphysic successfully meets these issues we must have a clearer understanding of what it is to be committed to brute tensed facts and what it is to provide an ontological underpinning for tensed truths. That is the aim of this chapter.

Here is a currently popular story. It has its origins in W. V. Quine,[1] was developed by David Lewis,[2] and finds its foremost contemporary defender in Theodore Sider,[3] with the metaphysical implications of the story getting more heavyweight with each next step in that chain.[4] There are differences in detail between the three, of course, and I shall focus particularly on Sider's account, both since he gives the most detailed account and because I am happy to view these issues with the metaphysical seriousness he himself views them with.

So here is the view. Theoretical commitment comes in two varieties: ontological and ideological. A theory's ontological commitments are the things it says exist: as Quine would have it, the things that must be in the domain of the quantifier if the theory is to be true.[5] A theory's ideological commitments are the primitive notions the theory utilizes: those notions which the theory makes use of in saying how things are which go undefined.

Put thus, ideological commitment sounds somewhat lightweight when compared to ontological commitment. It sounds as if we are talking merely about our *concepts*. So consider Lewis's attempt to avoid any commitment to ideology that

[1] See esp. Quine (1951). [2] See esp. Lewis (1983, 1986).

[3] See esp. Sider (2013a, 2013b).

[4] The pragmatist Quine would have no truck with, e.g., Sider's (2013a) talk of ideology tracking joints in fundamental reality.

[5] Quine (1980).

is modal.[6] Lewis wants to give an analysis of the modal concepts in non-modal terms. So we have the non-modal concepts of an individual, of mereological summation, of spatio-temporal relatedness, and of assessing a claim with respect to a restricted domain of quantification, and we then define the concept *is possible* as *is true when our domain is restricted to some maximal mereological sum of spatio-temporally related individuals*, with the other modal concepts being defined in terms of *is possible* in the familiar manner. There is nothing *wrong* with this description of Lewis's project, but it makes it easy to miss that Lewis is not aiming to say something about our mental lives—or at least, not *just* that—but that he is aiming to say something about *reality*. Lewis's view is that *reality* is a non-modal place, not (merely) that our modal concepts can be defined in terms of non-modal concepts.

What does it mean for *reality* to be a modal place? It cannot mean that there are modal *truths*, for of course Lewis thinks that there *are* modal truths: it is (strictly and literally) true, thinks Lewis, that 2+2 could not have been 5, and (strictly and literally) true that there could have been fewer planets in our solar system than there actually are. In Sider's terms, what it is for reality to be modal is for it to have modal *structure*. "Structure," for Sider, is a primitive notion, but it is easy to see what is intended. Moral structure is what the moral realist believes reality has but a moral quasi-realist like Blackburn does not believe it has.[7] The real realist and the quasi-realist cannot be differentiated by what representations they think correctly represent reality, for they both think claims like "It is wrong to torture babies for fun" are true. What differentiates them is what they think the truth of that representation is sensitive to. For the quasi-realist, the explanation for the truth of that claim resides ultimately in our attitudes, whereas for the moral realist the truth of moral claims is sensitive to something worldly. "Moral structure" denotes that in reality which the truth of moral claims is sensitive to, according to the (non-reductivist) moral realist. Note that this *may* be some thing or things: it may be a realm of moral *properties* or moral *states of affairs*. But it need not be: you could be a moral realist without thinking that there is any portion of ontology—any collection of *things*—to which the truth of moral claims is sensitive. You would still believe that reality has moral *structure*.

There is, of course, room for skepticism about the very idea of "structure." That is no surprise—there is room for skepticism about everything (or at least, for every thing, there is room for skepticism about it). And the quasi-realist might be one of the skeptics, pointing out that she herself can make all the same claims about moral structure in reality, just as she mimics realist claims about things

[6] Lewis (1986). [7] See the essays in Blackburn (1993).

having moral properties, about the obtaining of moral facts, etc. But we realists ought not to be concerned. We know what we believe in that the quasi-realist does not believe in, even if she does not. The fact that everything that we say can be deviously reinterpreted and assented to by the quasi-realist means that they are clever, albeit annoying, conversational partners, but it should not lead us to doubt that there is a distinctive realist position that we believe that is in conflict with the quasi-realist's anti-realist position. Berkeley will assent to all the ordinary claims we make about tables and chairs; with a little bit of devious interpretation the Berkeleyan can even happily assent to the claim that such things are made of mind-independent matter, etc. Nonetheless, we realists know what we believe—we believe that there *really are* mind-independent tables and chairs. The Berkeleyan does not, even if he will make the same utterances. I am happy to grant the intelligibility of structure talk, and will not engage further with a skeptic about the very coherence of the notion.

The Lewisian position on modality, viewed through this lens, is that reality lacks modal *structure*. Many modal claims are *true*, but their truth is not sensitive to something modal in reality, but rather to a certain non-modal structure in reality—namely, to reality's ontological structure (that is, what individuals there are) and to reality's spatio-temporal structure (in particular, the pattern of spatio-temporal (or analogous) relations that holds amongst the individuals). Lewis is giving an account of how modal *claims* can be true in an amodal world, that is, how the world can be correctly represented in modal terms despite the lack of modal structure for such representations to be sensitive to.

Seen in this light, ideology is as much worldly as ontology, and no longer looks like a lightweight commitment in comparison to ontological commitment. As Sider puts it:

> A fundamental theory's ideology is as much a part of its representational content as its ontology, for it represents the world as having structure corresponding to its primitive expressions. And the world according to an ideologically bloated theory has a vastly more complex structure than the world according to an ideologically leaner theory; such complexity is not to be posited lightly.[8]

> The term "ideology"... misleadingly suggests that ideology is about ideas—about *us*. This in turn obscures the fact that the confirmation of a theory confirms its ideological choices and hence supports beliefs about structure. A theory's ideology is as much a part of its worldly contents as its ontology.[9]

Ideology is not about us. And the relevant sense of "primitive" at play when we talk about our theory containing primitive modal (e.g.) notions is nothing to do

[8] Sider (2013a, pviii). [9] Sider (2013a, p13).

with our cognitive lives. The issue is not whether *is possible* can be conceptually analyzed in non-modal terms in the way that *is a bachelor* can be conceptually analyzed as *is an unmarried man*. Rather, the issue is whether any modal notion is—for want of a better term—*metaphysically* primitive: whether we need to deploy such a notion to give an account of the nature of reality, or whether such notions are ultimately dispensable. It is perfectly consistent with the success of the Lewisian project that some modal notion is *conceptually* primitive: that only tells you something about how our minds work, it tells you nothing about whether reality is modal, and Lewis is concerned with the latter. Again to quote Sider:

> [W]hether a property, word, or concept carves at the joints [i.e. tracks structure in reality] has nothing to do with the place of the concept in human languages, conceptual schemes, biology, or anything like that.... [It] signifies a metaphysical, rather than conceptual, sort of fundamentality. Humans may need to acquire other concepts first before they grasp joint-carving ones [i.e. structure-tracking ones]; and conversely, those concepts we acquire first, or most easily, may fail to carve at the joints [i.e. may fail to track structure in reality].[10]

Sider's position is that we have reason to believe that the primitive ideology our best theory utilizes to describe reality tracks the structure of reality. We should believe that reality contains modal/moral/epistemic, etc. structure if and only if our best theory of reality needs to utilize primitive modal/moral/epistemic notions in its description of what reality is like, just as we should believe that Xs are part of reality's ontology if and only if our best theory of reality needs to quantify over Xs in its description of what reality is like. He says:

> Quine's advice for forming *ontological* beliefs is familiar: believe the ontology of your best theory....We should believe generally what good theories say; so if a good theory makes an ontological claim, we should believe it. The ontological claim took part in a theoretical success, and therefore inherits a borrowed luster; it merits our belief....Its ideology is also more likely to carve at the joints. For the conceptual decisions made in adopting that theory—and not just the theory's ontology—were vindicated; those conceptual decisions also took part in a theoretical success, and also inherit a borrowed luster. So we can add to the Quinean advice: regard the ideology of your best theory as carving at the joints. We have defeasible reason to believe that the conceptual decisions of successful theories correspond to something real: reality's structure.[11]

And just as Quine tells us that it is a theoretical virtue to minimize our ontological commitments—that is, on his view, to minimize the number of (or number of kinds of) things in the domain over which your quantifiers must

[10] Sider (2013a, p5). [11] Sider (2013a, p12).

range if your theory of reality is to be true—so does Sider tell us that it is a theoretical virtue to minimize our ideological commitments—that is, on his view, to minimize the number of (or, perhaps, number of kinds of) primitive notions we make use of in our theory of reality. He says:

> Good theories must be as simple as possible, and part of simplicity is having a minimal ideology.... A theory with a more complex ideology posits a fuller, more complex, world, a world with more structure. Thus ideological posits are no free lunch.[12]

That leaves some questions open, of course. Does ontological parsimony trump ideological parsimony or vice versa or neither? Lewis seemed to think that ideological parsimony trumped ontological parsimony: when justifying his postulation of the vast realm of concrete spacetimes that his modal realism countenances he says that doing so "offers an improvement in what Quine calls ideology, paid for in the coin of ontology," and says that it is "an offer you can't refuse."[13] Why can we not refuse it? As ontological prices go, it's a big one[14]—maybe the price is too high. Lewis seems to be saying that obtaining ideological benefits is more important than avoiding ontological costs: that is, that facilitating ideological reduction is more important than not offending against ontological parsimony—that ideological parsimony trumps ontological parsimony. Personally, I think the most plausible view is that neither type of parsimony trumps the other, and that in cases where they pull against each other there is no general recipe for what we ought to do; we simply have to take each case as it comes and weigh up the pros and cons. But nothing will hang on this in what follows: all we will assume going forward is that a commitment to primitive ideology is *some* sort of cost—that ideological posits, just like ontological posits, need to be "paid for" by doing some theoretical work that could not be done otherwise (or at least, that could only be done at greater cost).

On the Quine–Lewis–Sider way of looking at things, what it is for your theory to commit you to tensed structure—for it to commit you to reality's being a fundamentally tensed place—is for your theory to resort to unanalyzed tensed notions when saying how things are. And given the above claims about theoretical virtuousness, such a theory will thereby incur a *pro tanto* cost over any theory of

[12] Sider (2013a, p14). [13] Lewis (1986, p4).

[14] Earlier (Lewis, 1973, p87) Lewis had made the claim that in fact there is no ontological price: that one only incurs an ontological cost if one increases the number of kinds of thing one believes in, but to merely believe in more things of a kind with what you already believe in—as modal realism demands (since everyone already believes in one concrete world, and it demands only that you believe in many)—is no ontological cost. But by the time of *On the Plurality of Worlds* he no longer makes this claim, retreating to the more plausible claim that modal realism does indeed pay an ontological cost, but that it is one worth paying.

reality that does not employ such unanalyzed tensed notions. Seen through that lens, it is easy to see why you would conclude that A-Theorists of every stripe must commit themselves to some kind of primitive tense ideology, for every A-Theorist thinks that reality used to be different, and lacks the resources, seemingly, to give a complete account of what it is for reality to have been different in un-tensed terms.

The presentist obviously thinks that reality used to be different. She thinks that there are not any dinosaurs—but there *were*. Hence she needs to invoke some tensed notion like "It was the case that" that she can apply to the false claim "There are dinosaurs" to yield a true claim concerning how reality used to be. Not believing in dinosaurs, the presentist *a fortiori* does not believe in dinosaurs that are before current goings on, thus she cannot follow the B-Theorist in saying that what it takes for there to have been dinosaurs is for there to be dinosaurs before current goings on and so, seemingly, she has to take "It was the case that" as a primitive ideological notion.

Likewise with the growing blocker. She thinks that there are both dinosaurs and humans. But there didn't used to be: that part of the block that includes dinosaurs but not humans used to be all there is. The growing blocker needs to account for this truth about reality: that the block used to be smaller. When it comes to "There were dinosaurs," the growing blocker says this is true because there are dinosaurs embedded in the middle of the block, but clearly the same story cannot be told when it comes to "The block used to be smaller": the block is not smaller anywhere in the block—no matter where you look in the block, the block is the size it is. How the block *is* is not going to help you account for how it used to be; as with the presentist, the growing blocker must avail themselves of some primitive ideological tense notion.

What of the moving spotlighter? She thinks there are dinosaurs, that there always have been dinosaurs, and that there always will be dinosaurs: reality does not change with respect to ontological claims. Nor, on the most obvious way of developing the moving spotlight,[15] do things change with respect to their ordinary properties. Of course, things will vary across time—but only in the same way the B-Theorist thinks they do: perhaps they have different temporal parts that are different ways, or perhaps they bear the redness relation to one time and the blueness relation to another time, etc. But, however that variation across time is cashed out, the resulting facts will not change: that temporal part always was and always will be red, that thing always did and always will bear the redness relation to time t, etc. However, reality as a whole is still subject to change: the spotlight

[15] But as we will see in the next chapter, I will not develop the moving spotlight view in this obvious way.

used to be somewhere different. As with the growing blocker accounting for the fact that the block used to be smaller, the fact that the spotlight used to be at a different time cannot be accounted for in terms of what is going on in reality in the bits the spotlight is not shining on, for no matter where in reality you look, the spotlight is where it is (i.e. on *this* time). And so the moving spotlighter is committed to some primitive ideological tense notion.[16]

So each of the main three versions of the A-Theory is committed—on the Quine–Lewis–Sider view—to taking as primitive some bit of tensed ideology. But to return to the ad hoc-ness threat that ended section 2.3, it is easy to think that the moving spotlighter has a particularly unsatisfying metaphysic here. The presentist puts their primitive tense to a massive amount of work: it lets us express the fact that there used to be dinosaurs, that I used to be shorter than I am now, that Caesar crossed the Rubicon, that there will be colonies on Mars, that I will have more gray hairs than I do now, etc, etc. (Even if the presentist wants to deny that claims like the last two are ever true, on the grounds that the future is open, she will still want to be able to *express* such claims, if only to say that they are false, or lacking in a truth-value, etc.) And it lets her say such things while keeping her ontology minimal: while avoiding ontological commitment to dinosaurs, Martian colonies, etc. Admitting primitive ideology might be costly, but the presentist at least can make a good case that this bit of primitive ideology is working hard to earn its keep.

But in believing in everything that the B-Theorist believes in and simply adding to that metaphysic a fact about what is objectively present,[17] the moving spotlighter seemingly has all the resources the B-Theorist has to give an account in tenseless terms of the truth-conditions of almost all tensed claims. When it comes to "There were dinosaurs," "Caesar crossed the Rubicon," "There will be colonies on Mars," and "I will have more gray hairs than I do now," the moving spotlighter can, seemingly, account for the truth or falsity of these claims by pointing to goings on that are before or after the spotlight. The primitive tense, seemingly, only needs to get invoked when we start saying that these goings on used to be or will be *present*. One worries both that this is ad hoc (shouldn't all tensed claims be treated alike?), and that it is not enough work to warrant the ideological cost. Here is Sider making exactly that latter complaint:

> Describing this change—the "motion of the spotlight"—is, in fact, the only purpose of the tense operators in the spotlight theory.... [The moving spotlighter] accepts that there

[16] Cf. Sider (2003a, pp94–5).

[17] Again, this is an obvious way of thinking about the moving spotlight. But as we will see in the next chapter, this is not how I will develop the view.

> exist dinosaurs before 2011; but this is the [B-Theorist's] proposed truth-condition for an utterance in 2011 of "There were dinosaurs." More generally, the spotlight theorist can accept the [B-Theorist's] reduction of tense for all tensed statements except those concerning presentness. For the latter, there can be no reduction...
> So the spotlight theorist can say *something* to justify her primitivism about the tense operators [i.e. that it allows her to say that the spotlight used to shine elsewhen]. But one might wonder whether it's convincing. The tense operators play no role in her theory other than characterizing change in the possession of the monadic property of presentness. And the only reason for invoking this property at all is to be able to say that there is genuine change in which moment is present. But notice that the spotlight theorist does not admit genuine change for anything else! For her there is no genuine change in whether I am sitting, or in whether there are dinosaurs, or in whether a war is occurring, since her account of these matters is identical to the [B-Theorist's]. All that genuinely changes is which moment has presentness. Is securing this smidgen of genuine change worth the postulation of primitive tense?[18]

This is a serious challenge for the moving spotlighter. Here is another way of thinking about it. Any A-Theorist is motivated by the thought that there is *genuine* change, not mere variation in how things are from one time to another. Surely, the thought here is that there is genuine change in the concrete goings on around us: in things like what height I am, whether there are dinosaurs, etc. That is the world we experience; that is how we come to learn about change. We do not start off theorizing thinking "You know, I'm really convinced that some other time used to have the property of presentness." That is a highly post-theoretic thought. The change we want to account for when we start theorizing does not concern times and presentness, it concerns dinosaurs, people, tables, etc. and their properties. If we cannot account for change in such ordinary, concrete goings on, we have lost track of our subject matter.

The B-Theorist offers an analysis of change in ordinary, concrete goings on: it amounts simply to variation across a dimension. The A-Theorist is unimpressed: surely, there is more to change than *that*? To account for *genuine* change, the A-Theorist invokes some primitive tensed ideology. There is *genuine* change when you need to invoke these primitive tensed notions, she tells us. The presentist takes this primitive tensed notion and uses it to tell us that the ordinary concrete goings on have indeed changed: there are no dinosaurs, but there *were*; I am not 5ft tall, but I *was*. Invoking the notion is a cost, but at least it secures the A-Theorist's guiding thought that there is genuine change in the concrete goings on around us. But the moving spotlighter looks like they have the worst of both worlds. When it comes to the concrete goings on around us, she thinks things are

[18] Sider (2013a, pp259–60).

just like the B-Theorist thinks they are. She only invokes her costly ideological primitive to make the highly post-theoretic claim that some other time had the property of presentness. The motivations for being an A-Theorist in the first place are not being served, it seems. The kind of change we care about is getting treated in the way that, by the A-Theorist's lights, renders it ultimately static—not *genuine* change. The only genuine change concerns something only a philosopher would care about. So there is a real worry here that the moving spotlighter is not finding adequate work for the ideological posit to do to make it earn its keep. If you are going to admit primitive tensed ideology, at least let it secure genuine change in the things we care about—otherwise why bother being an A-Theorist in the first place?

Now, of course, there are ways to be a moving spotlighter and avoid this complaint. Here is one view. In order to deal with the problem of temporary intrinsics, we accept that for a thing to be red (e.g.) at time t is for it to bear the redness relation to t. Except when t is the time on which the spotlight falls: to be red at the spotlit time is to instantiate, simpliciter, the monadic intrinsic property *being red*. You avoid the problem of temporary intrinsics because nothing ever has incompatible monadic properties, since the only temporary monadic properties you have (simpliciter) are the ones you have now. And on this metaphysic, there is genuine change in the things you care about. There is genuine change in what height I am. I am 6ft tall simpliciter; I also bear the *being 4ft* relation to a past time, but *being 6ft* is the only height property I have *simpliciter*. But this fact has changed: I used to be 4ft tall simpliciter; that is, it used to be the case that *being 4ft* was the only height property I had simpliciter, and I merely bore the *being 6ft* relation to the time at which I am writing this.

This version of the moving spotlight view avoids the charge that it invokes a costly ideological primitive only to secure change in matters unrelated to what was motivating adoption of an A-Theory in the first place, for on this view where the spotlight falls has an effect on the concrete goings on, thus genuine change concerning the spotlight *does* secure genuine change in what matters. However, it is not a very plausible metaphysic. For starters, a defender of this view needs to say why a red thing in the past and a red thing in the present are similar. The natural answer—that they are both red—will not do, for on this metaphysic it is simply not the case that both things are the same way: the thing in the past stands in a certain relation to that past time, and the thing in the present has a certain monadic property. We can *call* them the *being red* relation and the *being red* property, but that is just labeling. What *makes* them alike—why do they *deserve* the similar names? The answer could just be: the thing that bears the *being red* relation to t is like the thing that is red simpliciter because *when* the spotlight shone on t it *was* red simpliciter. But if that is the end of the story, what is the point of saying that the thing in the past bears this relation to that past time? Why not just cut out the middle-man and say that the thing in the past was red but is now no way with respect to color? And in fact, once we go down this road, why are we bothering to admit the existence of the thing in the past—why not simply say that it did exist and was red, but is now no more? We come back to a threat we raised earlier: letting your tensed notion do little work makes it look like it is not earning its keep, but once you start letting it do some serious work, the question becomes: why not go all in and simply be a presentist?

Here is another worry about this metaphysic. We want to know *how* the spotlight can have this consequence on the nature of concrete objects. How can a matter seemingly external to me—whether a time I am located at has the property of presentness bestowed on it in virtue of the spotlight's shining on it—have an effect on my intrinsic nature—whether or not I am 6ft tall simpliciter or merely bear the *being 6ft* relation to a certain time? Now maybe asking this question is to take the metaphor of the spotlight too seriously. Fine—but at the least, the defender of this view needs to tell us what is behind the metaphor that renders such a worry innocuous.

Let us take stock. Things are not looking great for the moving spotlighter. Like all A-Theorists she thinks reality used to be different, and does not think that you can say what it is (in full generality) for reality to have been different in un-tensed terms. Thus, on the Quine–Lewis–Sider view at least, the moving spotlighter, like other A-Theorists, is committed to primitive tense ideology. But the moving spotlighter's commitment looks considerably worse than the presentist's for at least four reasons:

(i) The work they put this new bit of ideology to is very minimal, far less than what it does for the presentist, and is not obviously worth the cost to ideological parsimony.

(ii) If the ideology they invoke could do all the work the presentist makes it do, this threatens to render the moving spotlighter's extra ontology redundant, thus undercutting one of their main apparent advantages over the presentist.

(iii) The work done by the new ideology seems remote from the concerns that were motivating the A-Theorist in the first place. The only change it lets us account for is change in what A-properties are had by things. Ordinary change—change in what there is, or in what ordinary properties things have—gets treated in the allegedly static manner the B-Theorist treats it.

(iv) The last point also raises the worry that the moving spotlighter has an ad hoc theory of change: while the presentist thinks that all change is a matter of something's being the case but it being true that it will or did fail to be the case, and the B-Theorist thinks that all change is a matter of something's being the case with your attention restricted to the present time (i.e. the time at which *this thought* is being had) but it not being the case with your attention restricted to some time that is before or after the present time, the moving spotlighter seemingly has to adopt a hybrid account. Some changes—such as that there are no dinosaurs but there used to be—get treated as the B-Theorist treats them: restrict your attention to the present time and there are no dinosaurs, but restrict your attention to some past time and there are. Other changes—such as that the spotlight used to shine on a time at which there are dinosaurs—get treated as the presentist treats them: the spotlight does not shine on such a time, but it is simply true of reality that it used to do so.

Reasons (i)–(iv) are real threats, and the moving spotlighter had better have something satisfying to say in response. What I will argue is that some of these threats only arise—or only look as bad as they do—when we view things through the Quine–Lewis–Sider lens. In the remainder of this chapter I will advocate a different way of thinking about things, and then in the next chapter I will develop a moving spotlight metaphysic which, with the meta-metaphysics about to be defended, has satisfying answers to these worries.

3.2 The Truthmaker Alternative

On the view described in the previous section, theories tell us two different types of thing about the world: what exists, and how reality is structured. This in turn leads to two theoretical virtues to be looked for in a theory: ontological parsimony, or the minimizing of ontological posits, and ideological parsimony, or the minimizing of structural posits. On the view I will defend, theories simply tell us what is true, and the only theoretical virtue concerning parsimony is to minimize the number of (kinds of) fundamental truths.

I take as primitive the notion of some proposition's being the case *in virtue of* some other proposition being the case. For p to be true in virtue of q is for q's being the case to explain, in more fundamental terms, why p is the case: that is, that p's being the case is a derivative feature of reality, whose obtaining is explained by the more fundamental feature of reality that q is the case. If we have a map of what holds in virtue of what, then we have an account of why the world is a certain way in terms of it being some more fundamental way. The fundamental (or brute) truths (if such there be)—those which are true and which do not obtain in virtue of any further truths—give us the ultimate foundation for the whole of reality: they characterize the world at its most basic level. The fundamental truths explain, in the most basic and most metaphysically perspicuous way possible, why anything that is non-fundamentally the case is the case.

Both this view—which is more in the spirit of that defended by Kit Fine[19]—and Sider's allow us to say of two truths that in some sense describe the same feature of reality that one does so better—that is, that it does so in a more metaphysically perspicuous manner. The difference lies in whether the features that make one truth more perspicuous than the other lie at the propositional/sentential level or at the sub-propositional/sub-sentential level.[20] If p and q are both true but p describes the same portion of reality that q does, but in a more metaphysically perspicuous manner, then for Sider that is because the sub-propositional components of p are structural—they all carve reality at its joints—whereas at least some of the sub-propositional components of q are not. So, for example, suppose Lewisian modal realism is true, and let q be the proposition that there could be blue swans and p be the proposition that there is a maximal mereological sum of spatio-temporally related individuals that has a blue swan as a part. These describe the same feature of reality according to Lewisian realism, but p does so better, and for Sider that is so because (if Lewisian realism is true) its sub-propositional components—the existential quantifier, the relation of spatio-temporal relatedness, the operation of mereological summation—are structural: they get at reality's joints. Whereas q has at least one sub-propositional component—the modal operator—that is not structural, since reality (assuming Lewisian realism) is a non-modal place. On the Finean view that I prefer, by contrast, if p and q truly describe the same feature of reality but p does so better than q, that is because p is more fundamental than q: that is, that q obtains in virtue of p.[21] So for example, suppose Divine Command Theory (DCT) is true. Then it is true both that torture is wrong and that God forbids torture, and both claims in some sense describe the same feature of reality; but the proposition that God forbids torture describes that feature of reality in a more

[19] See esp. Fine (2012).

[20] Sider (2013a, p128). It won't matter to what follows whether we talk in terms of propositions and their sub-propositional components or in terms of sentences and their sub-sentential components. I shall choose the former, but those happier with the latter can feel free to translate.

[21] This is a necessary but not sufficient condition for p and q to describe the same feature of reality. <There is an electron or there are round squares> is true in virtue of <There is an electron>, but these do not describe the same feature of reality, for the latter is more specific.

metaphysically perspicuous manner because it is the more fundamental truth (given DCT)—it is (necessarily) this in virtue of which it is true that torture is wrong. Here there is nothing sub-propositional to point to in order to say why <God forbids torture>[22] does better than <Torture is wrong> at describing reality: it is simply that one proposition explains why the other is the case.

To my mind, talk of one thing's being the case in virtue of something else's being the case is far more intuitive than talk of whether something sub-propositional, like a quantifier or a logical connective, is structural. The reason for going in for talk of fundamentality is to capture the thought that some truths only describe the world at a derivative level, and that there are more fundamental truths to be gotten at: it *sounds* like a relation that holds between something truth-apt, like propositions. Of course, this could be outweighed if the sub-propositional approach was more virtuous in other ways, and Sider and Fine have had a very interesting back-and-forth on the pros and cons of these two accounts.[23] But I am not going to get into that debate here; I am simply going to assume the more Finean view, and insofar as this book contains an argument for it, it will be the attractiveness of the metaphysic that is built on its foundation.

On the Finean view, in putting forth a theory of the world we will be (amongst other things) putting forth a claim about what the brute truths are (even if the claim is simply that there are none—that every truth obtains in virtue of some more fundamental truth). A *parsimonious* theory, on this view, will be one that has a minimal class of brute truths: that is, one that minimizes the number of, and/or the number of kinds of, truths that are taken to be true without being true in virtue of anything.[24] There are not two distinct theoretical virtues of ideological and ontological parsimony; ontological claims are just one kind of claim you can make about the world, along with modal claims, tensed claims, predicational claims, etc.—and parsimony tells us not to proliferate the (kinds of) claims that we take to be brute.

I accept a particular version of the Finean view, one that does very well by the above criterion of theory choice: namely, truthmaker theory. Truthmaker theory, as I understand it,[25] is a claim about what the brute truths are like: it says that only truths about what there is are brute. That is, the fundamental account of the world consists in simply specifying the ontological inventory, and every other

[22] I follow the standard convention of using "<p>" to refer to the proposition that p.

[23] Fine (2013) and Sider (2013b).

[24] As such, a theory on which there are no brute truths is maximally parsimonious in this respect. But this does not mean that we should believe in infinite descent, of course, for that might bring with it outweighing costs. See Cameron (2008a).

[25] See Cameron (forthcoming).

truth about how things are is true in virtue of the ontological inventory being as it is. This does very well on (qualitative) parsimony: only one kind of truth—ontological truths—it taken to be brute, as opposed to views that, in addition to taking as brute some claims about what exists, take as brute truths about what used to exist, or what could exist, or what ought to be, or how the things that exist are, or etc. That makes truthmaker theory a *pro tanto* virtuous theory.[26]

This is a departure from the traditional understanding of truthmaker theory, which is as a theory of *what it is* to be true. Truthmaker theory, as I understand it, is not a theory of truth: it is not a theory of what it is to be true, it is solely a thesis concerning what truths are, as a matter of fact, brute. Nothing about the *nature* of truth justifies our acceptance of truthmaker theory. Our explanation for why a proposition is true and our theory of what it is for that proposition to be true are two separate things, and they need not be related. The truthmaker theorist says that every truth is true because of what there is, and this is simply silent on the question of *what it is* for a truth to be true in the first place. It is compatible with this claim about what reality is like at the fundamental level that there is nothing more to something's being true than that what it says is the case is the case. Truthmaker theory is compatible with a deflationist theory of truth.

Thinking of truthmaker theory this way—as a theory of what truths are brute rather than as a theory of truth—gives us more room to maneuver when encountering thorny problems like negative existentials. True negative existentials, intuitively, have no truthmakers. They are true not because of what there is but because of what there is not. That has led some self-described truthmaker theorists[27] to say that some but not all truths have truthmakers: so "positive" truths like <There are elephants> have truthmakers, but "negative" truths like <There are no unicorns> do not. But if truthmaker theory is meant to be a theory of *what it is* to be true, this move looks untenable. If *what it is* to be true is to be made true by some thing(s) . . . well, it is true that there are no unicorns, so it had better be made true by some thing(s). This has led Armstrong to accuse theorists who think that some but not all truths have truthmakers of accepting a "dualism" about truth, and demanding to know what their theory of truth is for the truths they claim lack truthmakers.[28] If one thinks that truthmaker theory is a theory of

[26] The claim is that all the brute truths are existence claims. I do not claim that all existence claims are brute truths. "There is one way to win this chess game" might be true, but I do not think it is brute.

[27] E.g. Mellor (2003).

[28] Armstrong (2006, p245). I made a similar complaint myself (Cameron, 2008b, p412), but that was before I saw the light and realized that the truthmaker theory was not a theory of what it is to be true after all.

what it is to be true and wants to avoid such a dualism concerning truth, then one must hold that absolutely every truth has a truthmaker; but the cost of that is to admit to one's ontology weird entities like a totality state of affairs[29] or the absence of unicorns:[30] that is, one must admit the existence of some kind of unicorn-excluder—some thing that could not exist were unicorns to exist. But if truthmaker theory is taken to be a theory about what truths are brute, not a theory of what it is to be true, we can avoid the postulation of weird ontology like absences or totality states of affairs without ending up with a dualism concerning truth. We can let a totality *truth* be brute. That is, we can have a theory that says that the fundamental truths concerning reality are that a exists, that b exists, that c exists... and that nothing else exists. Certainly there is no commitment to a dualism concerning truth here, because this is completely silent on what it is to be true; it is a theory of what is fundamentally the case, nothing more. Nor is there a postulation of anything like a totality state of affairs or an absence by this theory; this theory says that it is true, indeed fundamentally true, that nothing other than a, b, c,... exist, but it does not postulate some entity that could not exist were that ontological inventory to be expanded. And so this view, unlike standard truthmaker maximalism, is compatible with the plausible claim that the world's ontology could have been strictly greater than it actually is—that everything that actually exists could have existed but have been accompanied by some thing that does not actually exist. What this view *does* do is acknowledge a fundamental truth not of the form "x exists," for it acknowledges as fundamental a truth of the form "Nothing other than a, b, c,... exists." But I do not think this is a particularly costly acknowledgment, and I think the resulting theory still coheres with the motivations behind being a truthmaker theorist in the first place. Truths of that form are as much truths about what there is as truths of the form "a exists" are. The idea behind truthmaker theory is that the only brute truths are ones that characterize the ontology inventory—that is, the truths that say what there is. To say what there is we have to make a list—and then we have to say that we're done! The totality truth is not of a different *kind* to the rest of the truths the truthmaker theorist (as I am characterizing her) takes as brute: it is simply another truth that characterizes what there is—that determines the ontological inventory. Thus admitting that some such proposition is fundamentally true is no further cost to qualitative parsimony.[31]

[29] As in Armstrong (1997, ch.13). [30] As in Martin (1996).

[31] If you are unconvinced by this, not a lot is at stake. Even if the truthmaker theorist has to admit two distinct kinds of truth as fundamental—truths about what there is and truths about what there is not—this is still more parsimonious than admitting truths of those kinds as fundamental *as well as* truths about what there was, and what there could be, and what there ought to be, etc., etc.

Sider is unconvinced that truthmaker theory can do what I am claiming for it, namely, that it offers us a sparse class of fundamental truths from which all else is ultimately explained. He says:

Fineans and I allow fundamental facts of arbitrary logical forms: predicational, negative, quantificational, and so on.... The truthmaking theory of fundamentality... countenances far fewer fundamental facts. The only fundamental facts, on this view, are certain singular existential facts, facts of the form "x exists", where x is a truthmaker...

It is very difficult to abide by this restrictive conception of fundamentality. And in fact, truthmaker theorists in practice almost never abide by it.... What in fact happens is that by making ineliminable use of certain bits of ideology, they smuggle in fundamental facts beyond those allowed by their theory.

One smuggling route is through the *canonical names* given to truthmakers, such as "the state of affairs of grass's being green". These names are formed by means of distinctive ideology, "the state of affairs of Φ", which is a functor that turns a sentence into a singular term naming a state of affairs. This functor acts as a cache for smuggling unacknowledged fundamental terms.[32]

Think of the point this way. Suppose God hands you a collection of entities: Alexander, Buffy, Cordelia, Dawn..., and asks you to work out the rest. Are there helium molecules? Are there cities? Why did *Buffy the Vampire Slayer* end after season seven? You wouldn't have any idea how to respond.... But according to the entrenched truthmaker theorist, the fundamental facts consist just of facts citing the existence of entities. It's hard to see how all the complexity we experience could possibly be explained from that sparse basis.[33]

I read Sider's objection as a dilemma. When the truthmaker theorist is telling us how she takes reality to be at its most fundamental level, she is to give us a list of what there is. In doing so, she has to choose how to refer to the things she is going to list: does she use "canonical" names—names which perspicuously reveal the nature of those things and thereby explain *why* it is that they make true the rest of the truths about the world, or does she use any old names to refer to the items on her ontological inventory? If the latter, then it looks like the theory is just a bad one—completely unexplanatory. If the former, then what is doing the work in the explanation is that the entities on the list are appropriately described by the canonical names, and hence any ideology used in forming those names must be taken as fundamental, and this thereby smuggles in fundamental facts other than what the truthmaker theorist said she was committing to.

Compare the objection Sider is making here against truthmaker theory to an objection he makes elsewhere against Finean essentialism.[34] He says:

[The Finean essentialist] says that "something flows from the essence of *A*", without saying *how* it flows. We haven't been given a picture of the "innards" of *A* from which we

[32] Sider (2013a, p157). [33] Sider (2013a, p160). [34] See esp. Fine (1994).

can just "read off" the claim.... The initial examples [Fine] uses, such as that of Socrates and singleton-Socrates, suggest that claims about essence can be underwritten by a special sort of explanatory story. In the case of singleton-Socrates, the underlying explanatory story involves the generation of all sets by an operation of set-building.... [T]he essential properties of each set arise because of how that set is built. But the eventual locution Fine uses to talk about essence is so much more powerful: a general operator $P_\lambda\varphi$, where φ can be any sentence and λ is any list of things. We are given no hint of how to understand arbitrary claims stated using the general locution, in terms of the tamer kinds of essential facts.[35]

An initial worry about this: I am not so sure that we have the kind of explanation that Sider seeks even in the case of "It is of the essence of singleton-Socrates that it contain Socrates as a member." Do we really have a picture of the "innards" of the singleton from which we can "read off" that it essentially contains Socrates as a member? I doubt it, for two reasons. First, even if we have a picture of the innards of singleton-Socrates such that we can read off that it has Socrates as a member, where does the *essentialism* come from? If I ever have a picture of the "innards" of a thing, I surely have it with respect to a thing and its parts; so if I am given a picture of A as "the sum of B, C and D" then I can read off from this picture of A that it has, for example, B as a part. But mereological sums do not have their parts essentially, and so it would be a mistake to conclude from this reading off from the picture of A that it has B as a part that it *essentially* has B as a part. What is different in the case of singleton-Socrates that makes it not a mistake to read off from the picture we are given of it that it not only has Socrates as a member but that it *essentially* has Socrates as a member? How can learning about the innards of a thing ever be enough to tell you whether the innards are essential to the thing or not? Sider talks about how in the case of singleton-Socrates we are given a story about where the set comes from—how it is *built* from its members. But while some building relations, like that which gets you sets from members, result in a thing which essentially stands in that relationship to the things from which it is actually built, others, like that which gets you mereological sums from parts, result in a thing which stands in that relationship to the things from which it is actually built as a matter of mere accident. And so I cannot see that we have been given any explanation at all of the essentialist claim; the essentialism is simply being smuggled into the explanation, at one point or another.

My second worry is that I am suspicious about this talk of the "innards" of singleton-Socrates in the first place. "Singleton-Socrates" is just a label, of course,

[35] Sider (ms.).

and calling something that name no more determines that it has some intimate relationship to Socrates than calling something "Armstrong" determines that it has strong biceps. The notation "{Socrates}" encourages the thought that Socrates is something like a part of the singleton, but of course he is not: there is no sense in which Socrates is "in" the singleton. He is a member of the singleton—but this is to say only that a certain relation holds between Socrates and singleton-Socrates. Some relations that hold between A and B do so as a result of the essence of one or both of A and B, and others hold between them accidentally—what is the reason for thinking that membership is of the former kind? Now, I *do* think that membership is of the former kind: I believe that if S has A as a member then it is essential to S to have A as a member. But I do not think I learn this by first learning something about the "innards" of S, or by learning how S is built from (amongst other things, perhaps) A. It is the other way round: I have reason for believing an essentialist claim—that sets have their members essentially—and I use this to *inform* my picture of S as bearing an especially intimate relation to A. Insofar as I am justified in thinking of the innards of singleton-Socrates as involving Socrates, I am justified in this by my *prior* belief in set-membership being a peculiarly intimate relation: one that describes the essential nature of one of the relata.

But put those worries to the side. It is clear what Sider is after. When the essentialist says that "It is of the essence of A that p," and when the truthmaker theorist says that B makes it true that q (or as I would have it: that q is true in virtue of <B exists>), Sider demands a specification of what A and B *are* that reveals why those essentialist/truthmaking claims are true. Alexander might be Socrates' singleton, but nonetheless "It is of the essence of Socrates' singleton that it has Socrates as its sole member" is good (in some sense of "good") and "It is of the essence of Alexander that it has Socrates as its sole member" is bad (in some sense of "bad"). Buffy might be the state of affairs of the postbox being red, but nonetheless "<The postbox is red> is true in virtue of <The state of affairs of the postbox being red exists>" is good and "<The postbox is red> is true in virtue of <Buffy exists>" is bad. Our theory of reality, in order to be a good and explanatory one, had better include the *good* claims. So the truthmaker theorist had better say not (merely) that <The postbox is red> is true in virtue of <Buffy exists> but rather that it is true in virtue of <The state of affairs of the postbox being red exists>. It is this latter proposition that she must list as being among the brute truths; but that is to thereby "smuggle in fundamental facts beyond those allowed by their theory."[36]

But how, exactly, does recognizing as brute the proposition <The state of affairs of the postbox being red exists> smuggle in a fundamental fact beyond

[36] Sider (2013a, p157).

those allowed by truthmaker theory? Truthmaker theory was advertised as admitting as brute only truths about what there is. Isn't this a claim about what there is? It says that there is some thing: a certain state of affairs. What is this proposition doing other than simply making an ontological claim: a claim about what belongs in our ontological inventory? I think that Sider is relying on his own view about fundamentality in making this objection against truthmaker theory. He is taking the fact that we use the *expression* "The state of affairs" when stating the proposition we take to be brute as bringing with it some commitment with respect to fundamentality. But this is to focus on *sub-propositional* features of the proposition. That is fine by his own lights, but not by the lights of the truthmaker theorist I have in mind.

Look again at the way Sider makes his complaint against truthmaker theory. He says:

> [B]y making ineliminable use of certain bits of ideology, [truthmaker theorists] smuggle in fundamental facts beyond those allowed by their theory. One smuggling route is through the *canonical names* given to truthmakers, such as "the state of affairs of grass's being green". These names are formed by means of distinctive ideology, "the state of affairs of Φ", which is a functor that turns a sentence into a singular term naming a state of affairs. This functor acts as a cache for smuggling unacknowledged fundamental terms.[37]

The complaint that is advertised is that the truthmaker theorist smuggles in extra fundamental *facts*. But what we end up with, when the justification for this complaint is given, is that she has smuggled in fundamental *terms*. But the truthmaker theorist does not think that *terms* are fundamental or otherwise: it is facts and only facts that are the locus of fundamentality or derivativeness.[38] And Sider nowhere specifies a *fact* that he thinks the truthmaker theorist has to take as fundamental that is of a kind she claimed not to allow as brute.

Sider seems to be arguing that, because we need to use certain language to *state* the fundamental truths in a good way (i.e. so that the resulting theory will be explanatory), we need to take that language (or perhaps the sub-propositional features denoted by that language) as fundamental. But that is to rely on his own meta-metaphysical approach, which sees as fundamental the bits of language we need to use in order to state our best theory of the world. The truthmaker theorist, by contrast, thinks it is true propositions (facts) that are fundamental, and whether or not you take as fundamental the proposition that the state of

[37] Sider (2013a, p157).

[38] Here and throughout, I am using "fact" to mean "true proposition," not an Armstrongian "state of affairs." (I take it this is how Sider is using it in the quoted passage as well.)

affairs of the postbox being red exists or the proposition that Buffy exists,[39] either way what you are saying about how things are fundamentally is just that a certain thing exists.

And to my mind, nothing hangs on the choice here. Either way we would be saying the same thing about the fundamental nature of reality: that a particular thing (Buffy/the state of affairs of the postbox being red) exists. The sense in which one gets a "better" theory by saying that the state of affairs exists is merely epistemic. It does not do you any good to be told that Buffy exists if you do not *know* what Buffy is, whereas you can work out what is being said to exist by the theory if I tell you that it says that the state of affairs of the postbox being red exists. But that does not mean that the *theory* is any better, only that this is a better way of stating it, given the facts concerning our competencies with various bits of language. If you are going to present me with the theory of evolution, it will go a lot better for me if you present it in English than if you present it in Aramaic, but that is simply due to my relationship to the representations, and says nothing about the theory or theories that are thereby represented. If I want to explain why the postbox is the color it is, the theory that says that Buffy exists is a perfectly good one. It specifies a way for reality to be at its most fundamental level such that this ensures that the postbox is the color it is. Of course, you can only *know* that this claim about how fundamental reality is ensures this claim about the color of postboxes if you know what Buffy is. But that is no surprise: to know whether a theory is a good one, you have to understand what it is that it is saying. The language used to present a theory can aid or hinder our understanding of what the theory says, but this affects only whether we know that the theory being presented is a good one, not whether it is in fact good.

Here is, I think, the crux of the disagreement between myself and Sider. Sider thinks that the fact that Buffy exists is no explanation for the fact that the postbox is red. The fact that Buffy exists *together* with the fact that Buffy is the state of affairs of the postbox being red might explain the fact that the postbox is red, but then there is a claim that is essential to our explanation of how things are that is not merely existential, for <Buffy is the state of affairs of the postbox being red> is not a claim about what there is, but a claim about what some thing that exists is like. In that case, the fact that the state of affairs of the postbox being red exists can only explain the fact that the postbox is red, thinks Sider, if it is taken to

[39] Of course, these might be the same proposition, given that Buffy simply is that state of affairs. But I will assume they are distinct propositions, to make life as difficult for myself as possible. Obviously if they are the same proposition, there can be no sense in which taking "one" of them as brute has consequences that taking the "other" as brute does not.

encompass both that the thing exists and that it is the kind of thing it is. But then that is to "smuggle in" as fundamental a fact not merely about what there is but about what those things are like—what kind of thing they are—and this violates the truthmaker theorist's self-imposed restriction. By contrast, I think that the fact that Buffy exists is, by itself, a perfectly good and complete explanation of the fact that the postbox is red. It is simply that we do not know that this is a good explanation unless we know that Buffy is the state of affairs of the postbox being red. But this does not mean that this fact about what Buffy is becomes part of the original explanans. The truthmaker theorist's claim is that the mere existence of A makes it true that p. A manages this because of what it is; the truth of p is ensured by the existence of A because of the nature of A. But the sole explanation for why p is the case is simply that A exists, not that A exists and has a certain nature. After all, *that* A has that nature is again simply made true by A.

3.3 On Giving an Ontological Underpinning

On the Quine–Lewis–Sider view, if one wants to keep expressing Φ-truths but wants to avoid commitment to Φ-facts being fundamental, one had better say *what it is* for a Φ-truth to obtain using Φ-free vocabulary. For example, if you are committed to there being truths about what is merely possible and what is necessary but you do not think that modality is fundamental, you had better be able to say, as Lewis does, what it is to be possible/necessary in non-modal terms.

The truthmaker approach defended in the previous section gives us another way of providing an ontological account for Φ-facts other than saying *what it is* for some Φ-fact to obtain in non-Φ terms: instead, we provide the *truthmakers* for Φ-facts—we provide an ontology that includes things from whose essence flow the Φ-facts. On the truthmaker approach, this is still to avoid commitment to a Φ-fact being fundamental, for what is fundamental is only that the truthmakers for the Φ-facts exist. Here are some examples to illustrate the two approaches coming apart.

Consider the attempt to ground modal truths in the powers of individuals.[40] On this view, every modal truth can be explained by the various powers things have. It is possible that I have a sibling rather than being an only child. Why? Because my parents have the power to procreate more than once and, had that power been exercised, I would have had a sibling. It is possible that this intact glass be smashed. Why? Because the powers of this hammer and this glass, were they exercised, would result in a smashed glass. Now personally, I am

[40] See e.g. Jacobs (2010), Vetter (forthcoming).

unconvinced that every modal truth can be explained in this way,[41] but put that issue aside. Let us grant that every modal truth has an explanation that cites solely the powers of existing individuals; what would that achieve? Certainly not a reduction of the modal to the non-modal in the Quine–Lewis–Sider sense. For one thing, the notion of a power is itself clearly a modal notion. For another, since some powers might always go unexercised, in saying why something is possible we need to speak about what *would* happen *were* some powers exercised, which is explicitly modal. So the powers theorist is not offering an account of *what it is* to be possible/necessary in non-modal terms. But what their theory *does* do is say from where possibility arises. It provides an ontological underpinning for modal truths by providing in each case a truthmaker for the modal truth in question (whether that truthmaker be the state of affairs of a thing having the power, or merely the individual that has the power itself, will depend on whether you think a thing has its powers essentially). That means, if the view defended in the previous section is right, there are no fundamental modal facts, for the fundamental facts are simply that these things (the individuals that have the powers, or the state of affairs of those individuals having those powers) *exist*, and it is because they exist that the resulting modal claims are true.

Or consider the attempt to analyze predication. If that means saying *what it is* for n individuals to stand in an n-place relation (with monadic properties being understood as 1-place relations) then, plausibly, this cannot be done. We could say that what it is for A to be F is for instantiation to hold between A and F-ness, and in general that what it is for n things to stand in an n-adic relation G-ness is for instantiation to hold between those n things and G-ness. But this leaves at least one type of predicational fact—facts about instantiation itself—unaccounted for. Saying that what it is for instantiation to hold between A and F-ness is for instantiation to hold between A, F-ness, and instantiation itself might well be true, but it only tells you what it is for A to instantiate F-ness by invoking the very notion we are trying to account for. Thus, on the Quine–Lewis–Sider approach at least, there is no way to avoid "primitive predication." Indeed, Lewis uses precisely this fact to conclude that no theory can avoid primitive predication, thus aiming to undercut some of the objections to nominalism.[42]

But as Daniel Nolan has argued,[43] the truthmaker theorist can provide an ontological account of predication in a different sense. For each predicational fact, she can say what *makes it true*, thus accounting for whence predicational facts arise. On a states of affairs ontology of the kind proposed by David

[41] See Cameron (2008c, p273), Wang (forthcoming).
[42] Lewis (1983).
[43] Nolan (2008).

Armstrong,[44] for every predicational fact there is a truthmaker for that fact. What makes it true that A is F? The state of affairs of A being F. What makes it true that instantiation holds between A and F-ness? Again, the state of affairs of A being F. An ontology of states of affairs does not allow us to reductively say *what it is* for a predicational fact to obtain, since states of affairs are states of affairs of things *being* a certain way, and that "being" just *is* the predicational tie. But they do allow us to resist taking any predicational fact as fundamental, for what is fundamental is simply that the states of affairs exist: the fundamental facts are solely ontological facts, concerning what there is, and do not include any facts concerning things *being* a certain way. It is due to the nature of the things whose existence is claimed by those fundamental facts that things are a certain way.

A final example. Some people have the resources to say *what it is* for one thing to be a part of another in non-mereological terms, namely, those who believe that composition is identity.[45] If composition is identity, then what it is for A to be a part of B is for A to be among some things that are (collectively) identical to B. But to the majority of people who reject this doctrine, the prospects of saying in non-mereological terms what it is for one thing to be a part of another are dim; this is why van Inwagen concentrates on saying *when* composition occurs rather than what composition *is*.[46]

But despite not being able to say in reductive terms what it is for one thing to be a part of another, or what it is for some things to compose some thing, we might still be able to provide truthmakers for each mereological fact. Consider a metaphysic whereby it is a fundamental fact that certain simple things exist, but where unrestricted composition holds and where, for every object O that is a sum of some simple Xs, the fact that O exists and has the parts it has is made true simply by the fact that the Xs exist. This is a metaphysic on which it is of the essence of each thing to compose with whatever other things there may be: it is part of the nature of every individual to enter into the composition relation with whatever else there may be. Now again, believing such a metaphysic does not furnish one with the resources to say *what it is* for some things to compose some further thing. It is not true that what it is for the Xs to compose some thing is for them to exist. The existence of the Xs and the fact that they compose some thing are two distinct features of the world: it is just that they necessarily go along together, and one feature is prior to the other, just like the existence of Socrates and the existence of Socrates' singleton are two distinct features of the world that

[44] Armstrong (1997).
[45] See Sider (2007b), Wallace (2011), and the essays in Baxter and Cotnoir (2014).
[46] van Inwagen (1995).

necessarily go along together, with one being dependent on the other. But while the believer in this metaphysic cannot reductively say what it is to compose, she can say what makes it the case that things compose—it is due to the nature of the simple things that exist. And she need take no fact about composition as fundamental—what is fundamental is simply that certain simple things exist, and it is entirely due to this feature of the world that the world is one in which composition occurs.

Now let us return to tense. The B-Theorist offers an account of what it is for an A-Theoretical fact to obtain in B-Theoretic terms: what it is for something to be present just is for it to occur simultaneously with some event that is stipulated to be present in this context (such as my having *this* thought), and what it is to be past/future just is to be before/after that which is present. The A-Theorist cannot do this: there will be at least one tensed fact such that she cannot say what it is for that fact to obtain without appealing to some A-Theoretic notion. But this does not prevent her from offering an ontological underpinning for all tensed facts: she can still provide *truthmakers* for every A-Theoretic fact, and thus avoid having to take any tensed fact as fundamental.

Operating within the Quine–Lewis–Sider framework, the moving spotlight did not look like a very good theory. Given that it fails to provide an account of what it is for every A-Theoretic fact to obtain (since it has to take as primitive the fact that the spotlight used to and will be elsewhere), it is not clear why the extra ontology over a presentist metaphysic is warranted. What I will aim to show in the next chapter is that while the moving spotlighter's extra ontology does not allow them to reductively say what it is for a tensed fact to obtain, it *does* provide them an advantage over the presentist in providing *truthmakers* for each tensed fact. So going forward, the challenge will be to provide a metaphysic whereby no tensed fact need be taken as fundamental—that the obtaining of each tensed fact be explained by the existence of some things. This will be the task of the next chapter.

4

The Moving Spotlight

In this chapter I present my moving spotlight metaphysic and argue that it meets the desiderata and overcomes the challenges raised in the previous chapters. A central idea is that we ought not to think of the moving spotlight as taking the B-Theorist's metaphysic and *adding* to it that one time is objectively privileged. Instead of thinking of the moving spotlight as an enriched B-Theory, we should think of it as an enriched presentism: on the moving spotlight view I defend, the presentist is right that how things are, simpliciter, is how things are *now*. There is no reality at all to how things merely were or will be. The difference between presentism and the moving spotlight theory as I develop it is simply that on the latter, non-present as well as present entities are some way *now*. This means that in pursuing the truthmaking project from chapter three, we must posit things with rich enough natures such that each tensed truth gets made true by things being the way they are now. In §4.2–§4.3 (having made explicit some assumptions of the project in §4.1), I develop an account of the natures of things at a time that is rich enough for them to speak to how they were and will be at other times. In §4.4 I argue for the moving spotlight theory over presentism on the grounds that the truthmaking project goes better if we can appeal to non-present entities as being amongst the things that are now a certain way. In §4.5 I argue that any endurantist must accept the claim that my version of the moving spotlight theory and presentism agree on: that the only way things are, simpliciter, is the way they are now. In §4.6 I answer a challenge to the effect that my view cannot uphold an obvious principle of tense logic. In §4.7 I argue that my proposed metaphysic meets the desiderata and overcomes the challenges that were revealed in the previous chapters. I end in §4.8 by providing a summary of the main theses of the moving spotlight view that has been defended.

4.1 Goals and Assumptions

In this chapter I will present the moving spotlight metaphysic that I claim meets the desiderata and overcomes the challenges that have been brought out in the previous chapters. The goal will be to put forward a metaphysic whereby there is a truthmaker for every tensed fact; while we will not be demanding a reductive account of *what it is* for things to have been a certain way (etc.), we will be demanding that for every truth about how things were, will be, or are now, there is something in the postulated metaphysic to make that truth true.

Call the claims that concern the temporal goings on in the world the "temporal claims." A good metaphysic of time needs to be able to account for the truth-value of all the temporal claims at every time. I say this to narrow our subject matter: a complete metaphysic ought to say what makes it the case that 2+2 is 4 (even if the answer is: nothing), but clearly it is no demand on the success of a metaphysic of *time* that it account for the mathematical truths, which simply do not concern temporal goings on. Similarly, suppose you are a non-Humean about the laws of nature, and you further hold that what laws of nature obtain is contingent but fixed: that is, that it vary from world to world what laws hold, but that in no possible world does it vary from time to time what laws hold. In that case, while there is certainly an interesting question concerning what makes it the case that the actual laws of nature obtain, it is not the kind of question that we should demand our metaphysic of time to answer: this question simply does not concern the temporal goings on of the world.

It is sufficient for p to be a temporal claim that it be subject to change. This is compatible with p's truth-value in fact never changing; all that is required is that, as a result of its subject matter, it is the type of claim that *could* change in truth-value: there is a metaphysically possible world in which it is true at one time and false at another. So, for example, perhaps it is possible that I change gender but in fact will not do so. In that case, even though my gender never in fact changes, our metaphysic had better be able to say what makes it the case that I will always be the gender I currently am. It could have changed, so we need an account of why it does not do so. But it is not necessary for p to be a temporal claim that it be subject to change: "The Universe started with the Big Bang" is a contingent truth that, necessarily, is always true or always false. Nevertheless, it is clearly about the temporal goings on in the world, and so is a temporal claim. I will not attempt a definition of "temporal claim"—I think it is clear enough what claims count and what do not.

Any metaphysic of time had better be able to account for the truth-values at every time of every temporal claim. In our case, for every temporal claim, p, our moving spotlight metaphysic had better be able to say what makes it true/false now that p, or what makes it the case that p was/will be true/false when time t was/will be present, for all past and future times, t. Here is an assumption I will make going forward:

> *Supervenience of Temporal Claims*: The facts concerning the truth-values of temporal claims at different times supervene on the facts concerning the locations and intrinsic natures of things at different times.

Supervenience of Temporal Claims tells us that once you have fixed the facts concerning where things are and how things are intrinsically at every time, you have thereby settled all the temporal claims.[1] *Supervenience of Temporal Claims* rules out a situation whereby some temporal claim changes but where this change in truth-value is completely untethered to a change in the intrinsic natures and/or locations of some things. So suppose you were a non-Humean about the laws of nature: you thought that which laws obtain was not fixed by the facts concerning how ordinary, concrete things are from one time to another. Suppose you are led by this to the view that there is simply a brute fact concerning which laws of nature obtain, but you nevertheless allow that this fact is subject to change. This is the kind of view that is ruled out by *Supervenience of Temporal Claims*. If this is your view, you will allow that there could be two possible worlds exactly alike with respect to how and where things are at each time, but which differ with respect to which laws hold at each time. For example, there could be a world w in which there are two particles whose movements are governed by law L up until time t, and where after t their movements are governed by law L*, but where their movements after t are nonetheless consistent with the previous laws, and another world v where the same two particles have the exact same history but where the laws L obtain throughout. The difference between w and v is whether their movements after t are mandated by the laws (as they are in w) or whether they are merely permitted by the laws (as they are in v). w and v differ concerning the truth-value of some temporal claim at some time—'L obtains' is a temporal claim that is true after t in v but false after t in w—and yet they do not differ concerning the location and intrinsic nature of what exists from one time to another; since these two particles are the only things that exist in either world, they move in the exact same ways in each world, and (we can assume) they undergo no intrinsic change in either world. So this is the kind of situation that is ruled out by *Supervenience of Temporal Claims*: to hold this supervenience thesis is to hold (*inter alia*) that the laws of nature are not like this. Either what laws obtain from one time to another depends on the intrinsic nature and location of things from one time to another, or the facts concerning what laws obtain are not subject to change.

I find *Supervenience of Temporal Claims* plausible. But really, I am assuming it just to make life easier: to narrow the class of things we have to worry about when

[1] Actually, really what I am assuming is something stronger: that once you have made true all the facts concerning how things are intrinsically and their locations at each time, your ontology thereby has all you need to *make true* every fact concerning the truth-value of a temporal claim at some time. But I will focus on the supervenience claim for simplicity; nothing will hang on this.

putting forward a metaphysic of time. If you hold a view on which *Supervenience of Temporal Claims* is false, the metaphysic I am going to offer is not incompatible with that, but it will be incomplete. I am going to be offering an account of how all the facts concerning how things are intrinsically at each time, and where they are at each time, get made true. If *Supervenience of Temporal Claims* is true, then once I have done that my job is done. If it is false, then there will be some tensed truths that have as yet gone without truthmakers. And so if you think *Supervenience of Temporal Claims* is false, I invite you to supplement my offered metaphysic with an account of how the missing tensed truths get made true.

Here is another assumption I will make going forward:

> *The Primacy of Present Instantiation*: the way things are intrinsically now is the way they are intrinsically simpliciter.

Now, that might look like a strange principle for a moving spotlighter to endorse. Isn't the point of non-presentist metaphysics like the moving spotlight to admit the reality of how things were and/or will be? In saying that how things are intrinsically simpliciter is how they are intrinsically now, are we not denying the reality of their past/future intrinsic natures? Is it not the *presentist* who, in thinking that reality simply does not extend beyond the now, should think that a thing's intrinsic nature, simpliciter, is exhausted by its intrinsic nature now?

The presentist should certainly think that. But in fact I think that every A-Theorist should take a leaf out of the presentist's book here. In fact, every A-Theorist should hold something even stronger, namely:

> *The Primacy of Present Truth*: What is the case now is what is the case simpliciter.[2]

As I see things, accepting *The Primacy of Present Truth* is *what it is* to accept that the present is privileged in the way the A-Theorist wants. What used to and will be the case just isn't any part of the way things are simpliciter. The temptation for the non-presentist A-Theorist to deny *The Primacy of Present Truth* arises, I think, from the thought that we need to recognize the reality, simpliciter, of how things were/will be, and so there is more to how things are simpliciter than how things are now. But as we saw in chapter two, it cannot be the case in full generality at least that how things were/will be is a way things are simpliciter. And

[2] This is to accept the characteristic tense logical principle that all presentists will accept: NOW (p) iff p.

I think that once we have given up on that claim in full generality, we should just abandon it entirely: how things were/will be is merely a way that reality was/will be, and it is no part at all of how reality is simpliciter.

To accept *The Primacy of Present Truth* is to get away from the idea that the moving spotlight metaphysic simply takes the B-Theorist's metaphysic and adds to it an additional feature: that some time be objectively present. Thinking of the moving spotlight metaphysic that way really does suggest that how things are now is but a fraction of how things are simpliciter: that how things are simpliciter is how the B-Theorist thinks they are, but with one extra fact, whereas how things are now is how things are with one's attention restricted to the special time. But as we have seen in the previous chapters, this is not how the moving spotlighter should think of things: she should not think of her view as being simply the B-Theory with the added extra of one moment being objectively present, for then she cannot distinguish her view from the stuck spotlight metaphysic with a fancy semantics that lets her talk as if time passes, and it threatens to render the property of being objectively present a mere metaphysical idler.

So all A-Theorists, presentist and non-presentist alike, should accept *The Primacy of Present Truth* and *The Primacy of Present Instantiation*. The difference between the presentist and non-presentist A-Theorist, as I see it, is not whether they admit more to how reality is simpliciter than how reality is now, but rather whether they allow that non-present things are nevertheless now a certain way.

The moving spotlight metaphysic I will offer in this chapter accepts the presentist's thesis that what is the case simpliciter is what is *now* the case. What merely was or will be the case is accorded no reality at all on this moving spotlight metaphysic, and this is how the moving spotlighter will avoid McTaggart's paradox. It also means that to account for truths about how things were or will be, the presentist and moving spotlighter alike must say that these are made true by things *now* being a certain way (since how things are now is the only way things are, simpliciter). But one reason for being a moving spotlighter rather than a presentist, I will argue, is that this truthmaking work is easier if we allow that some non-present things are nonetheless now some way or other. The presentist, believing that there are no non-present things, can only allow this if she allows that non-existent things can nonetheless be a certain way. But the moving spotlighter can allow this without dabbling in such Meinongian-esque deviant metaphysics, for she of course thinks that there *are* non-present things. She denies that they are located at the present time—but why should that stop them from now being some way or other?

For the presentist "A is not present" means that A is not now amongst the things that there unrestrictedly are: A is not to be found in the realm of being. So since dinosaurs are not present, for example, the presentist thinks our ontology is lacking in dinosaurs. Thus, to allow that some dinosaur nonetheless now is a certain way is to deny that being is a necessary precondition for being a certain way. That principle is highly intuitive, so the presentist ought instead to deny that some non-present object is now some way or other. The moving spotlighter, by contrast, thinks that "A is not present" simply says something about A's *location* in time: that A is not located at the time that is objectively privileged. Whether or not A is within the realm of being is not something that is subject to change: if A ever exists, it is always true that A exists.[3] So "A exists" is now true if it ever was. The non-presentness of dinosaurs simply amounts to a claim about the location of dinosaurs: none of them are located at the present time. But for every dinosaur that ever existed, it is true now that that dinosaur exists. So there is no problem in the moving spotlighter saying that each of those dinosaurs now is some way or other, and in particular that they each now have a certain intrinsic nature. There is no force to the principle that being located at the present time is a necessary precondition for being a certain way, which is the principle the moving spotlighter has to deny.[4]

My moving spotlighter does not admit non-present entities because she thinks she must admit the reality of how things were or will be—she has learned from chapter two to reject *Past Record* (and its future analog). Rather, she admits non-present entities because it helps with the project of making true tensed truths by appealing to things now being a certain way: that project is made easier by

[3] Assuming that "A" is a concrete substance. See the remarks in the introduction.

[4] Saying what I want to say can be a bit tricky, because both "A now exists" and "A is located in the present" are ambiguous. "Caesar now exists" can mean that it is now true that Caesar exists, which of course the moving spotlighter accepts and the presentist denies. But it can also be heard as saying that Caesar exists *in* the present—a claim about Caesar's location in time—which both the presentist and the moving spotlighter will deny. Similarly, "Caesar is located in the present" can mean this claim that both presentist and moving spotlighter deny, but it can also be heard as saying that it is now true that Caesar is located (somewhere or other), which the moving spotlighter will accept.

I trust that context will usually make it clear what I mean. But just to fix terminology going forward: I will only use "A is located in the present" to mean that A's location includes the present and never to mean simply that it is presently the case that A has *some* location. (So the moving spotlighter will deny "Caesar is located in the present," in the relevant sense.) And I will only use "A now exists" to make a claim about what is now true regarding the unrestricted realm of being, and never a claim about the temporal locations of those things. (So the moving spotlighter will think that no claim of that form ever changes in truth-value, so long as "A" refers to a substance rather than a state of affairs.) I will use "A is present" and "A is non-present" as a convenient way of making the locational claim: that A is/is not located in the present time.

being able to say that non-present entities are now a certain way, and that is only an acceptable thing to say if you think it is now true that there *are* non-present entities, which is incompatible with presentism. Doing this will not let the moving spotlighter say *what it is* for something to have been the case in the past, but it will let her say what *makes it the case* that each historical truth was the case—which, following chapter three, is all we want.

4.2 Against Lucretianism

Assume for simplicity that nothing goes in or out of existence. (We will dispense with this assumption in section 4.4.) In that case, given *Supervenience of Temporal Claims*, in order to provide a grounding for all the tensed truths, all we need to do is provide truthmakers for truths concerning the past, present, and future intrinsic nature and location of each of the present things. Let us ignore location for the moment (again, we will come back to it in section 4.4): what we want is an account of how things are intrinsically now that is rich enough so that how things are intrinsically now settles how each of those things was and will be intrinsically at every past and future time.

An easy way to do that is to hold that any thing's intrinsic nature at a time includes tensed properties concerning how it was and will be intrinsically. This is Lucretianism, and has recently been defended by John Bigelow as a presentist solution to the truthmaking problem for historical truths.[5] The Lucretian says that the fact that I used to be 4ft tall is made true by my now having the property of *having been 4ft tall*.

Lucretianism has not proven very popular. Trenton Merricks complains that Lucretian properties like *having been 4ft tall* are irreducibly past-directed and hence "suspicious" and that "it is a cheat to rely on these properties."[6] Theodore Sider, similarly, complains that such properties "point beyond" their instances.[7] But, as Sider admits, this is a pretty elusive notion; it would be disappointing if there was nothing more we could say about exactly what is wrong with a metaphysic that deploys these properties.

Here is a first shot at doing just that: the property *having been 4ft tall* is suspicious because it points beyond its instances—is irreducibly past-directed—in the sense that a thing's presently having that property tells us nothing about how that thing is *now*, it only tells us about how that thing *was*.

That is not satisfying yet. Because we *are* told something about how a thing is now if we are told that it has that property: we are told that it is now such as to

[5] Bigelow (1996). [6] Merricks (2007, p135). [7] Sider (2003a, p185).

have been 4ft tall. We can complain that this is not the right kind of fact we want to find out about from knowing what properties an object currently has; but why not? Saying what is "wrong" about that fact is no easier a challenge than saying what is wrong with the property in question in the first place. We have not made an advance on the problem, merely postponed answering it.

Here is what I think is wrong with Lucretianism. In section 4.1 we said that we wanted an account of the intrinsic nature of each thing just now that was rich enough to settle how it used to be, and how it will be, intrinsically. What is wrong with the property *having been 4ft tall*, I suggest, is that it is *not* part of a thing's intrinsic nature *now* that it was 4ft tall: a thing's now having that property tells us nothing about how it is intrinsically *now*, it only tells us something about how it *was* intrinsically—namely, that it *did* have the intrinsic property *is 4ft tall*. This is the sense in which the property points beyond its instances: a thing's presently having that property tells us nothing about the present intrinsic nature of the thing.

I am relying on there being a sensible distinction between the intrinsic nature of a thing *at a time* and its intrinsic nature *across time*—that is, its *atemporal* intrinsic nature. An object's instantiating *having been 4ft tall* does indeed tell us something about the intrinsic nature of that object if by its intrinsic nature we mean its atemporal intrinsic nature; it just tells us nothing about its intrinsic nature *at the time the property is being had*. Here is a way of getting a grip on this distinction if it is not immediately intuitive. (This is not intended as an analysis, but rather as a rough guide on how to classify cases.) *Being 6ft tall* is part of the intrinsic nature of an object that has that property at the time it has it, whereas *having been 4ft tall* is merely part of the atemporal intrinsic nature of an object that has it; by contrast, *being the tallest thing* is not intrinsic in either sense. *Being the tallest thing* is not intrinsic at all because the thing that has it could have lacked it given a change in its surroundings only. Whereas for any thing that has the property *having been 4ft tall*, it could only have lacked that property had *it* itself been different at some time or another. However, this property is not intrinsic to its bearer *at the time it is instantiated* (hence it is only part of its atemporal intrinsic nature), because a thing's having this property depends on its having had a particular past. I have the property of *having been 4ft tall*, but I could be exactly as I am right now but with a different past and not have this property. Whereas *being 6ft tall* is part of my intrinsic nature *right now*. My having this property does not depend on my having a particular past (or future), so for me to lack this property there would have to be a change in how I am right now: for any possible past that I could have had, any way I could have been up until now, it is possible that I have had that past and presently be 6ft tall, and it is possible that I have had that past and presently am some other height. I have my extrinsic properties partly in virtue of how my surroundings are. I have my atemporal intrinsic properties in virtue of having had a certain history or future, but independently of my surroundings (both how my surroundings are now, and how they were and will be). I have my present intrinsic properties solely in virtue of how I am now, independently of both my surroundings and my particular history and future.

The suspiciousness of properties such as *having been 4ft tall* consists in their not making a contribution to the intrinsic nature of their bearers at the time at which they are instantiated. I suggest that this is what makes it "cheating" to rely on such properties when postulating truthmakers for tensed truths. The only properties we should admit into our sparse base class of properties are ones that I will call *difference-making*: ones that *make a difference* to the intrinsic natures of their bearers at the time they are instantiated. That is, ones that are part of the determination base for the present intrinsic nature of their bearers at the time of instantiation. That is:

> *Intrinsic Determination*: For all objects x and sparse properties F and times t, if x instantiates F at t, then x has the intrinsic nature at t that it has partly *in virtue of* instantiating F at t.

Combined with the further constraint that we should only appeal to sparse properties in providing truthmakers, this rules out Lucretianism. And this further constraint is reasonable. In providing truthmakers for a class of truths we are saying what those truths hold in virtue of. The sparse properties are the ones we ultimately appeal to when saying why things are the way they are: if something is F, we should be able to say in virtue of what it is F mentioning only its sparse properties. So if some truth p holds in virtue of some thing's being G, G must be a sparse property—otherwise, there would be a deeper explanation for the truth of p.

So every property we allow to do truthmaking work has to make a difference to how its bearers are *now*. But some of the truthmaking work we want to do is make true claims about how things were and will be. Hence, we must find properties that play a dual role: properties which both make a difference to the present intrinsic nature of their bearers—properties which are *difference making*—*and* which fix the truths concerning how the bearer was in the past and will be in the future—call these properties *history settling* properties (in the sense of "history" in which there can be a "future history"). My suggestion is that we appeal to *distributional properties*.

4.3 Distributional Properties

Distributional properties, as characterized by Josh Parsons,[8] are properties that say how a thing is *across* some region. Spatial distributional properties say how a thing is across a region of space. The property of *being polka dotted*, for example, says something about the pattern of the bearer across space. Something is not polka dotted at a point: it is polka dotted in virtue of being a certain way from point to point—that is, in having a certain *distribution* of color.

Consider an object that is white with black spots. That is how it is *across* space. Now, it is also true that the object is wholly white at places and wholly black at others. But how can that be? Nothing is both wholly white and wholly black. Does it have some spatial parts that are black simpliciter and others that are white simpliciter? Perhaps; but perhaps not—perhaps this object is an extended simple.[9] In order to account for that possibility, Parsons suggests that the object simply has a certain distributional property: being white with black spots (in a certain way).[10] Instantiating this property explains why it is the way it is across space; but it also, says Parsons, explains why it is the way it is at the subregions of space it occupies: it could not have the particular distribution of black on white it in fact has without being wholly black at its center (for example).

A *temporal* distributional property says how a thing is across time just like *is polka dotted* says something about how a thing is across space. Consider a simple world consisting of just one spatial dimension and one temporal dimension. There is one entity in this world—Flatty—which starts off its life at time t as a point, but who as time progresses grows continuously[11] in one direction of the one spatial dimension it occupies. After the beginning of this life, then, it is no longer a point but a line; and at each moment it is a longer line than it has ever been previously. Time t* is the last moment of Flatty's life: after this time, Flatty no longer exists, and the world is empty. At t Flatty is a point, at t* it is a line. But how can this be? Nothing is both a point and a line. What we should say is that Flatty is neither a point *nor* a line: rather, Flatty is (atemporal "is") a

[8] Parsons (2000, 2004).

[9] I will argue below (see fn 32) that qualitatively heterogeneous extended simples are in fact impossible. But it will still prove helpful to think about such things at this stage, to get a grip on the notion of a distributional property.

[10] It's hard to *state* what distributional properties a thing has, because we do not have words for the ways three-dimensional objects are across time or for most of the ways objects are across space. But we should not let this linguistic fact count at all against the legitimacy of such properties: what there is has little to do with the tools available to us to *describe* what there is.

[11] Flatty's length is a function of its age, and for every positive value of length less than the length it currently is, there is a past time at which it was that length.

triangle.[12] This is to attribute to Flatty a temporal distributional property that says that Flatty is a certain way *across a period of time*.

Flatty's having the temporal distributional property it has explains why it is the way it is across time. But with one other property, it also explains why it is the way it is *at* the particular moments in time it exists. The other property is its age: a property which says how far along Flatty is in its life. Fix its age and its temporal distributional property and you have fixed how Flatty *is*, how it *was*, and how it *will be*. Furthermore, both its age and its temporal distributional property contribute to Flatty's present intrinsic nature, since it is in virtue of instantiating them both that it is *now* intrinsically the way it is, as well as that it was/will be some intrinsic way.

Flatty could not have the particular distributional property and age that it has without being pointy at t and without being that length of line at t*, and it is *because* it instantiates this distributional property and age that it is pointy at t and that length of line at t*. And so the existence of the state of affairs of Flatty instantiating this distributional property and age at t* necessitates that Flatty was pointy at t, and can suffice as a truthmaker for that truth about the past. And this entity is not suspicious in the way the state of affairs of my being such as to have been 4ft tall is suspicious. Instantiating this distributional property and age at t* *does* make a difference to the intrinsic nature of Flatty at t*. It is in virtue of instantiating these properties at t* that Flatty is a line at t*, and this is part of Flatty's intrinsic nature at t*. So I think temporal distributional properties and ages play the dual role we were looking for: they are difference making, for it is in virtue of having both that the bearer is the way it is intrinsically *now*, but they are also history settling, since it is in virtue of having those same properties that the bearer did and will have the intrinsic nature it had/will have at other times.

It is *not* true that Flatty could not have been that length of line at t* had it not had that very temporal distributional property and age. There are infinitely many combinations of ages and temporal distributional properties it could have had which would have resulted in it being that length of line at t*. So intrinsic nature

[12] Why is Flatty a triangle? At no time does Flatty look triangular, of course, but if we think of Flatty *across* time then it looks triangular. Flatty's world has one dimension of space, and one dimension of time; imagine drawing Flatty on a two-dimensional grid, with the spatial dimension of Flatty's world as the vertical axis and the temporal dimension as the horizontal axis, so that as you move along to the right of the grid you are representing Flatty at a later time, and a line going up the grid represents what size of line Flatty is at the appropriate time. If you draw Flatty on this grid as it is specified in the text to grow across time, then you will have drawn a triangle. So when we think of Flatty across time, as we should when we are thinking about its distributional properties, we should say that it is a triangle. At every time, it is a line (a line of zero length at the start, which is a point)—but across time it is a triangle.

at a time is multiply realizable. Also, nothing I have said precludes (although it might render it unwarranted) that Flatty *also* (e.g.) bears a *being of length l* relation to t* and that it is that length of line at t* in virtue of bearing this relation to t* *as well as* in virtue of instantiating at t* the distributional property and age it instantiates. So intrinsic nature at a time might be overdetermined. Furthermore, necessarily everything has *some* temporal distributional property and age or other, so you cannot take Flatty's temporal distributional property and/or age away from it without giving it another temporal distributional property and/or age: one which will also make a difference to its intrinsic nature at the time of instantiation.

For these reasons, no modal analysis of "difference making properties" will work. One might have initially been tempted to say that a difference making property is one such that any object that has that property at t could not have lacked it at t without having had a different intrinsic nature at t. That is, that the difference making properties are those that you cannot take away from a thing and leave its present intrinsic nature the same. But this will not work, precisely because if you take away an object's temporal distributional property you must give it another one in compensation, and it might make the same contribution to the object's *present* intrinsic nature (while having made a different contribution to its past intrinsic nature, perhaps) that the one you took away did. Nor can we say that a difference making property is one where taking it away *might* change the present intrinsic nature of the bearer. *Having been 4ft tall* will pass that test; but that is not because something's having it is making a difference to its present intrinsic nature, but rather because if you take it away from something you must also take away the temporal distributional property the thing has in virtue of which it was 4ft tall, which *is* making a difference. So you cannot take away this non-difference making property without also taking away a difference making property: and so it is hopeless to try and analyze "difference making" in modal terms.

For an illustration of the view, consider two worlds, w and v. One particular exists in both worlds: A. In w, A has just come into existence: it is presently F, but in the next instant[13] will be G and in the next instant after that H. In v, A came into existence an instant ago. It is presently G, but was F and in the next instant will be H. In both worlds, how A is across time is grounded by its instantiating the same temporal distributional property: *being-F-then-G-then-H*. But there is a difference in truth between the two worlds: in w A is *now* F and in v A is *now* G.

[13] Let's pretend time is discrete. This is just for ease of illustration; nothing in my account relies on time being discrete.

What is the difference in being between the two worlds? In w, A instantiates the age *being-newly-existent* and in v A instead instantiates the age *being-one-instant-old*. The facts concerning how A is, was, and will be are all settled by A having the age and distributional property it has. A cannot be one instant old and have this distributional property without being G and being such as to have been F and such that it will be H. And since A's having these properties is what grounds its present intrinsic nature—that it is *now* intrinsically G—these properties are not "peculiar" in the way the Lucretian's are.

I think the postulation of distributional properties and ages gives us an acceptable story about how facts concerning how present things used to and will be are made true by the way things are now.[14] And what is crucial to the account is that facts about the distributional properties or ages things had/will have are settled by the nature of the distributional properties and ages that things now have. Things do not change their temporal distributional properties: it is in virtue of A having the temporal distributional property it now has that it always had and always will have that very temporal distributional property. After all, a temporal distributional property is a property that says how a thing is across time, and one of the ways things are across time is that they do not vary in how they are across time as a whole. Things do change their age properties, but in a completely predictable manner: necessarily, if something has the property of *being one year old* now, then in a year's time it will have the property of *being two years old*, and it is in virtue of it now having the former property that it will then have the latter property. (If you are worried about what right I have to insist that things do not change with respect to their temporal distributional properties while allowing that they do change with respect to their ages, I address this in section 4.6.)

This is how the view avoids McTaggart's paradox. Consider a past event involving present entities, such as my being born. That event was present, so the threat of McTaggart's paradox is that it *is* present in the past which, since the past is the only place this event is, threatens to entail that it is present simpliciter. I deny that the event of my birth is present in the past (I deny *Past Record*). My birth *was* present, but its being present is simply no part of reality. That is a way the world was; it is not a way any part of the world—not even the past—*is*. The presentness of that event has been and gone—it is no longer to be found in reality.

[14] Should we *want* a story about how claims about how things *will be* are made true? Isn't granting that there are such facts about things in conflict with the thesis that the future is *open*? I think it is not, and in the next chapter I will argue that the metaphysic offered in this chapter is compatible with the open future.

But reality makes it the case that it *was* present: the way I am now makes it the case that I was the way I was when I was being born, and hence makes it the case that this event was present. In particular, my having the age I have now makes it the case that I used to be a younger age, and my having the temporal distributional property I have now makes it the case that I always have had that temporal distributional property. Together, then, they make it the case that I was the way I was at my birth, which is just to say that they make it the case that my being that way was present. The presentness of the past event need be no part of reality, for we have a rich enough present to make it the case that this is how things used to be.

Now, so far we have ignored facts about location, and we have assumed that nothing goes in or out of existence. We will need to complicate the story to deal with both, but before we do that let me respond to some potential objections to my use of temporal distributional properties.

Objection 1: "Your temporal distributional property is equivalent to a complex conjunctive property, some conjuncts of which are simple present-tensed properties like *is 6ft tall* and others of which are the past-directed properties like *was 4ft tall* that Sider and Merricks find peculiar. So you have not really avoided the peculiar past-directed properties after all."

Reply: What does "equivalent" mean? It is true that for any temporal distributional property, there is a very complex big conjunctive property made up of the objectionable past-directed properties and unobjectionable present-directed properties instantiation of which would have exactly the same effect as the distributional property. In that sense, they are equivalent. But (according to the view being put forward) the temporal distributional property is fundamental, whereas the conjunctive property is not. It is not true that a man instantiating this conjunctive property was 4ft tall in virtue of having that property, just as it is not true that he is now 6ft tall in virtue of having that property. He is now 6ft tall in virtue of having the particular conjunct *is 6ft tall* and he was a boy in virtue of having the other conjunct *was 4ft tall*—and so properties which are purely past-directed are doing truthmaking work, which is objectionable. The point of appealing to distributional properties is that this move cannot be made. Distributional properties cannot be broken up into simpler components: there is just one property here, and it is fundamental—and it is an instantiation of exactly the same property that is making true truths about how the bearer now is that is making true truths about how the bearer *was*. There is one fundamental property that is involved both in history settling and in difference making, and so there is no sense in which we are appealing to a solely past-directed property in our truthmaking.

Objection 2: "Sider's objection to Lucretianism is that those properties 'point beyond' their instances. What is bad about them is that they say something about how their bearers are at times other than the time they are being instantiated. That is still true of your solution: an object's instantiating one of these distributional properties now 'points beyond' itself in that it says how the object *was*. It does not matter if it *also* says something about how the object *is* (at the time of instantiation): the fact that it says something about how the object is at a time *other* than the time of instantiation is enough to make it susceptible to Sider's objection, and hence they are just as peculiar in this respect as the Lucretian's properties."

Reply: We can separate two different potential charges of "peculiarity": two ways of understanding what it is for a property to "point beyond" its instances. There is the charge that a property is peculiar because instantiation of it at time t entails something about the intrinsic nature of the bearer at some time t* such that t≠t*; and there is the charge that a property is peculiar because instantiation of it at time t entails nothing about the intrinsic nature of the bearer at time t. I have argued that whereas Lucretian properties are peculiar in the latter sense, distributional properties are not. However, distributional properties and Lucretian properties are indeed both peculiar in the former sense. But I do not recognize this sense of "peculiarity" as something that ought to be avoided. We should not rule out that some properties can only be had by an object if they are had by it across an extended period of time. Perhaps nothing could be charged without it being charged over a (perhaps arbitrarily small) extended period of time. Perhaps not; but my claim is just that we should not rule out cases like this. In that case, if something has charge at an instant it is also true that it either was charged at some previous instant or will be charged at some future instant. So an object's instantiating *being charged* at t entails something about its intrinsic nature at some times other than t: in this sense, the property "points beyond" its instances. But that just seems utterly unproblematic; what is objectionable is only if instantiating a property at t *does not* say anything about the intrinsic nature of the bearer at t. Lucretian properties are so objectionable, but distributional properties are not.

Objection 3: "What if things are infinitely old? What grounds the fact that some thing is the way it is now rather than the way it was a year ago (say)? Its age (*being infinitely old*) and distributional property are the same in both cases, so what accounts for the difference in intrinsic nature?"

Reply: Dealing with infinite time just requires us to get more creative about ages. Suppose time were infinite in the past but finite in the future. Then the age of things can be thought of as a countdown to their end rather than as the count from their beginning. And so things have a different age now than they had a year ago even if they are infinitely old: they are one year closer to their end. Suppose instead then that time is infinite in both directions. Well, just pick an arbitrary time—1980, for example—and think of the age of things as giving the distance from that time: an age of −10 years putting things 10 years before 1980 and an age of +10 years putting things 10 years after. Then time can be infinite in both directions and things still have different ages at different times. I do not think ages understood as distance from 1980 should be objectionable if ages in any sense are permissible, since the bog-standard conception of ages I have been working with in the above is basically equivalent to distance from the *first* time. Why should distance from the first time be acceptable and not distance from some other time if there simply happens to be no first one? (Do not think of this view as "'privileging" 1980 in some objectionable way. The time we pick really is arbitrary, and the picking of an arbitrary time is just a way for us to get a grasp on the age property. It makes no difference whether we think of age as distance from 1980 and describe the age of things in the year 2000 as *being 20 years after time t* or whether we think of age as distance from 2050 and think of the age of things in 2000 as *being 50 years before time t*. These are just two ways of thinking about exactly the same property.)

All that I need, when I am looking for properties that play the role of ages, is that there be a kind of property such that (i) every thing that exists in time has exactly one property of that kind at every moment at which it exists; (ii) no thing has the same property of that kind at two different times; (iii) the way in which an object changes with respect to properties of this kind is predictable in the following sense: that as a matter of necessity, for any thing a and any times t and t*, if a exists at t and t*, then a's having the property of this kind that it has at t, together with the temporal distance between t and t* and the fact about which time is earlier, settles what property of this kind a had/will have at t*.

If there is a finite past, then ages—thought of very intuitively, as the count of how long a thing has been around—fulfill this role. If there is an infinite past then we need to think of ages a bit differently, but the existence of properties fulfilling this role is not incompatible with the infinite past (or future). *Perhaps* there are recherché metaphysically possible worlds in which time has some really weird structure whereby nothing like this is going to work. But that is acceptable: I am claiming only that something like this *does* work, not that it *must*. I do not demand that the account offered here be necessarily true; I offer it as a true account of the actual world only.

I do not think it should be any constraint in responding to a truthmaker objection that the account offered be a necessary truth. For one thing, I do not think we have any reason to think that truthmaker theory is a necessary truth. The case I made for truthmaker theory in chapter three appealed to theoretical virtues: we should believe truthmaker theory, I suggested, because if it is true there is a sparse class of brute truths that are all of a kind—the truths that are just about what exists—and it is a theoretical benefit to have fewer brute truths, and fewer kinds of brute truth, than more, then there is a *pro tanto* reason to accept truthmaker theory over theories that take as brute not only truths about what there is but *also* truths about how those things *are*, how they *were*, how they *could be*, how they *should be*, etc.

But principles of theory choice like simplicity, parsimony, etc. are generally contingently reliable at best—they might have taken us badly wrong. And while there is no *inconsistency* in a contingently reliable principle of theory choice selecting a necessarily true theory, I have argued elsewhere that we should believe the theories selected by contingently reliable principles of theory choice to be themselves contingently true unless we have some independent reason to believe that they are necessary.[15] We have no such reason when it comes to truthmaker theory. It is theoretically beneficial to be able to explain everything in terms of what exists, but that suggests only that there are *in fact* no brute facts concerning how things were, not that there could not have been. There is no *inconsistency* in the postulation of brute truths other than existential truths, nor is truthmaker theory a consequence of any necessary truths concerning the nature of truth or realism.[16] The best we can do to justify truthmaker theory is to argue that it is beneficial if true; but since the world need not have cooperated with what is beneficial (we just hope it in fact does), this only gives us reason to believe in the truth of truthmaker theory, not in its necessity. Had time been so weirdly structured so as to make no account like mine work, perhaps there would have been brute past-directed truths. Or perhaps Lucretianism would have been true: after all, the reason we rejected Lucretianism was the peculiarity of properties that make no difference to the intrinsic nature of their bearers at the time of instantiation—but again, that only gives us reason to reject the actual existence of such properties, not their possible existence. Many peculiar things *could* have existed; we just hope the actual world is better behaved.

[15] In Cameron (2007, 2008a) I apply such reasoning to argue for the contingency of the facts concerning when composition occurs, and the claim that there is a fundamental level of reality, respectively.

[16] See Cameron (forthcoming).

4.4 Why We Need Non-present Entities

On the proffered metaphysic, all truths about how things were or will be get made true by things being a certain way *now*. In that sense, it looks like a presentist-friendly metaphysic. So why not just be a presentist? As I said above, I think believing in non-present things makes this truthmaking work easier, because we can say that non-present things nevertheless are now a certain way: you do not need to be *located* in the present in order to presently be a certain way, it simply needs to be now true that you exist.

So far we have ignored facts about a thing's location. But once we extend our account to such facts, the moving spotlighter is at an advantage over the presentist. The moving spotlighter can say that each thing bears a location relation to some four-dimensional region of spacetime; and this relational fact is one that always obtains, because the relation fixes not where the thing is *at* a time, but where it is *across* time. In that case, the way a thing is now—its bearing the location relation to a certain region of spacetime—makes it the case that it always has and always will be located in that region of spacetime, and this of course fixes what spatial region the thing occupies at each time. But the presentist cannot say this. She does not believe in other times, or in four-dimensional regions of spacetime, so the only locations she has for things to be located in are present spatial regions. The location relations things now stand in fix merely which of these present spatial regions they occupy. But because things move around in space (in a way that is not predictable: a thing's past or future spatial location is not fixed by its present spatial location), she needs to say what makes it the case that things used to and will be elsewhere. There are things the presentist can say, of course. She can, for example, build locational facts into the temporal distributional property. So instead of saying just that I have a temporal distributional property that describes the changes in my height, mass, etc. over time, it also describes the changes in my location over time. But this is unsatisfying. A's being located where it is is an extrinsic feature of A, and extrinsic features should have relational truthmakers. That is, A's extrinsic features should be made true not (solely) by A's having a certain monadic property but rather by A's standing in a certain relation to its surroundings (at least in part). Of course, whenever A stands in a relation to some things, the Xs, we can postulate a monadic property that A has iff it stands in that relation to the Xs. So, for example, if A stands in the *is the mother of* relation to each of the Xs, she also instantiates the property of *being the mother of each of the Xs*, and vice versa. But, I suggest, relational monadic properties like that should not be counted amongst the sparse properties that we ought to limit ourselves to when postulating truthmakers.

Facts about the extrinsic natures of things are made true by things standing in relations to other things, not by their having a relational monadic property that makes reference to the other relata. Facts about A's location should be made true by A's bearing the location relation to the region in which it is located, not by A's having a monadic locational property of *being in region R*. Now, the presentist can hold that "A was in spatial region R" is made true by a relational fact: the obtaining of the *was located in* relation between A and R. But for a relational fact to obtain, both relata must exist, so that means that R must be a spatial region that exists now. So for this story to work, there must be no change over time in what spatial regions exist, otherwise there could be a truth of the form "A was/will be in R" that cannot be made true, because R is not around now for A to stand in the *was/will be located in* relation to it. But what guarantee does the presentist have that regions of space cannot be annihilated or come into being? The presentist believes that things come into and go out of existence,[17] so why not the regions in which things are located? At the very least, she is leaving herself hostage to fortune to contingent matters of fact here in a way the moving spotlighter need not do.

The other main advantage for the moving spotlighter concerns changes in what there is. Attributing temporal distributional properties to things lets us describe the way they have varied their intrinsic qualitative nature across time. So for things that are around now to have such properties, this lets us say what makes it the case that there is a part of their life where they are F, a part where they are G, etc. And adding ages allows us to say which part of their life has been, which is still to come, and which is present. But what about the life of things that are no longer around? There are not now any dinosaurs, for example, so what makes it true that there were dinosaurs? I think the presentist—if she is to hold on to the principle that truths about how things were are made true by things now being a certain way—has two choices, neither of which is attractive. First, she can hold that while there are not now any dinosaurs there *are* now things that *were* dinosaurs. If there are now things that were dinosaurs then those things can now have a temporal distributional property that describes their early life as a dinosaur, and so this thing's now being the way it is makes it true that it used to be a dinosaur (hence that there were dinosaurs). This is to accept the temporal analog of Timothy Williamson's view about modality.[18] Williamson thinks that, necessarily, what there is exists necessarily. So there is no variation from one possible world to another as to what exists. So every human necessarily exists. But

[17] Unless she is a Williamsonian presentist, of the kind discussed in the following paragraphs.
[18] Williamson (2002, 2013).

clearly there is *some* sense in which I am a contingent existent: my parents might never have met and, had they not met, I would never have been born. Williamson agrees: he thinks that had that been the case, I would have existed in the "merely logical" sense, but I would not have existed as a concrete being. The temporal analog[19] says that the things that were dinosaurs still exist now, but they exist in the merely logical sense, not as concrete beings (and hence they are no longer dinosaurs, since dinosaurs are concrete). If the presentist takes this route, she can say that all facts about what there used to be (such as that there used to be dinosaurs) reduce to facts about how things that now exist used to be (such as that some merely logical existent used to be a dinosaur). But the cost is clear: this ontology, to my mind, is far less attractive than the moving spotlighter's. If the choice is between admitting the existence of dinosaurs in the past or admitting the existence of mere logical existents that used to be dinosaurs that exist now but that aren't located *anywhere*, I'll take the former. The presentist's other option[20] is to say that the world itself has a certain temporal distributional property that describes its overall history. Since one of the ways the world is across time involves it containing dinosaurs, the world's presently having the temporal distributional property and age it has can make it true that there were dinosaurs. Again, I think this is unsatisfying. Unless one is a priority monist like Jonathan Schaffer,[21] and thinks that the world is the most fundamental being, then surely the world has its intrinsic nature in virtue of the things that make it up being the way they are, not vice versa. The world has the history it has because there were dinosaurs, not the other way round. The other problem with this view is that we do not only need something to make true that there were dinosaurs, but we need something to make true that there were the *particular* dinosaurs there in fact were. It is not merely the qualitative facts about how things were that need to be made true, but also the facts about what individuals were around to *be* such-and-such a qualitative way. We can say what makes it true that a diplodocus existed; but we still need to say what makes it the case that Deborah the diplodocus existed, rather than her merely possible intrinsic duplicate. The Williamson-style proposal has no trouble here, because Deborah still exists, and how she is now makes it the case that she was a diplodocus. But if the temporal distributional property of the world itself is to make it the case that Deborah existed, then the temporal distributional properties cannot merely describe the *qualitative* variation a thing undergoes throughout its life. The world's temporal distributional

[19] Which Sullivan (2012) has defended.
[20] This is what I defended on the presentist's behalf in Cameron (2011).
[21] Schaffer (2010).

property, at least, must be about particular individuals having been part of the world. And this must be a *fundamental* way the world is, for it to do the truthmaking work required of it. To my mind, at least, this is unacceptable: the fundamental ways things are do not involve non-qualitative features of things other than themselves.[22]

The moving spotlighter, of course, faces no problems with any of this. There are dinosaurs located in the past, she thinks, and that makes it true that there were dinosaurs. Indeed, the dinosaurs located in the past are the very dinosaurs that once existed, so this also makes true the non-qualitative facts concerning what particular dinosaurs existed. And those dinosaurs now have a certain age and temporal distributional property that makes true all the facts about how they were at the various times at which they were present.

So my moving spotlighter thinks that everything—whether it be a past, present, or future entity—is *now* a certain way, and that this is how it is *simpliciter*. Those things were and will be a different way—which is to say, as time passes how every thing is,[23] simpliciter, changes. Past things continue to change as time progresses: Caesar is a certain age now, but will be a different age tomorrow. Things in the distant future will be different when they are in the not-so-distant future, and different again when they are present. A yet to be born sun has a certain age, will have a different age when it is born, and will have yet a different age when it dies. This change in the ages of things, combined with their having the temporal distributional property they always have, results in things changing with respect to their more mundane properties. I am 6ft tall because I have a certain temporal distributional property and a certain age, and I was 5ft tall because I used to have that same distributional property but a different age, and this fact about the properties I used to have is made true by my having the temporal distributional property and age I now have. Caesar used to be a certain height at time t1 and a different height at t2, and he had those heights because of the different ages he had at each time in combination with the same temporal distributional property; and

[22] What of, e.g., Socrates' singleton? Isn't a fundamental way Socrates' singleton is that it has some distinct thing, Socrates, as a member? I don't think so. What may be the case is that it is of the essence of Socrates' singleton to stand in a certain *relation*, the set membership relation, to something distinct from it, namely, Socrates, and the obtaining of this relational fact may be a fundamental way things are. That is fine; what is unacceptable is saying that something has a fundamental monadic property that involves the non-qualitative features of some thing distinct from it. I have no problem with someone saying that there is a fundamental relational fact relating the world and Deborah. But for there to be such a fundamental relational fact, the relata have to exist. So for the presentist to say this, she has to think that Deborah exists, which takes her back to the Williamsonian view.

[23] Every thing that is in time, at least. Entities that exist outside time, such as numbers, need not undergo any change at all as time progresses. (Numbers need not have ages.)

that he had those properties at those times is made true by his *now* having the temporal distributional property and age that he has—the properties he has *simpliciter*. And Caesar's now having these properties makes it the case that he now has no height—that he is neither 6ft tall, nor 5ft tall, nor . . . etc.—because his temporal distributional property describes him as having a height only for an initial portion of his life, and his age makes it the case that that portion is over.

It is a consequence of my view that Caesar exists now that he stands in a certain location relation now (and thereby is now a concrete individual), but that he now has no height, no mass, no ordinary 3D shape, etc. Indeed, the only properties Caesar now has are ones he always has (such as his temporal distributional property, his location to a certain region of spacetime, and his essential properties such as being concrete, being a human, etc.), and his age, as well as the properties that follow from these.[24] I admit that this is a cost, as the following conditional (*) is very intuitive:

> (*): If a concrete thing exists now, it must now be some height, some mass, some 3D shape, etc.

On reflection, however, it should be clear that not many people are in a position to push this as an objection to my view. The perdurantist is not, for she also rejects (*), since on her view many concrete individuals, such as you and me, are not the right type of thing to have heights, masses, 3D shapes, etc. It is only instantaneous temporal parts, not persisting objects, that have such properties. (Perhaps there is a derivative sense in which the perduring worms have mass, namely, the sum of the masses of each temporal part. But this is certainly not the normal sense of mass that is featured in the laws physicists talk about. And I can allow that Caesar now has mass in a similar derivative sense: he can have the sum of all the masses he had at each moment he was present.) Similarly, an eternalist endurantist who takes apparent temporary monadic properties to be, for example, disguised relations to times (see the next section) denies that there are such properties as being height h, or being mass m, etc. So she will reject (*) as not

[24] To tie this back to the discussion of Forrest's "dead past" hypothesis in chapter one, then, I agree with Forrest that Caesar is not conscious, and hence is simply not having any thoughts about being present. But unlike Forrest, I also deny that the neurons in Caesar's brain are firing a certain way, that he is related to the Rubicon a certain way, etc. Forrest holds that Caesar is the way he was in every respect *other* than being conscious. You might have thought that whether someone is conscious supervenes on what their brain is like, etc., but Forrest thinks it supervenes on this *and* their being present. Being present looks magical—it turns the non-conscious world into a conscious one without changing anything else! By contrast, I hold that Caesar is none of the ways that is relevant to a thing's being conscious: neurons are not firing in his brain, for example. They *were* firing, and he *was* conscious, but they and he are that way no longer.

well-formed on the grounds that it presupposes that things have such properties. She will consider as well-formed the alternative principle (**): if a concrete thing exists now, it must bear some height relation, and some mass relation, etc. to the present time. But she will certainly reject (**), for she holds that Caesar exists now (i.e. it is now true that Caesar is amongst the things that unrestrictedly exist), but bears no height relation to the present time. The one theorist who can accept (*) is the presentist, since she believes in properties such as being height h and being mass m, and in attributing these properties to ordinary persisting concrete objects she can allow that every concrete object that exists now, now has some such property. I will concede that the presentist has an advantage over me here; in securing (*) the presentist can legitimately claim to be upholding the common sense position. I simply claim that this is an advantage that is outweighed by the advantages listed in this section of admitting non-present entities.

What makes it true, on my moving spotlight view, that our time is present and that other times were/will be present? One could go different ways on this. One could say, for example, that there are things, times, that have temporal distributional properties and ages just like ordinary objects such as you and me, and that these temporal distributional properties describe times as being non-present, then non-present, then non-present, then..., etc...then present, then non-present, then non-present, then..., etc. Then combined with its age, this will entail whether a time is present, has been present, or will be present. I do not want to go this route because I do not want the fundamental properties to be about presentness. I want an explanation of *what it is* to be present. One of the potential problems that was raised for the moving spotlight view in the introduction was that, unlike presentism and the growing block view, it offers no good account of what a time's being present amounts to. This view does not offer a good answer, it merely postulates a primitive way for things to be with respect to presentness. Every theory needs its primitives, of course, but we should try to do better.

I want to say that the facts about the presentness of times are just reflections of the facts about how ordinary objects are. On my view, the global way things are, simpliciter, is never realized more than once. How all things are at any one time is never a way that all things are at any other time. If for no other reason, this is true simply because nothing ever has the same age at more than one time. In that case, there is a mapping from times to properties of things. In which case, to say that some time, t, is/was/will be present we simply need to account for why the things that there are have/had/will have the properties that they get assigned from t on that mapping.

What makes it true that noon (GMT) on January 1, 1980 was present? That simply amounts to the question: what makes it the case that everything was the way it was at noon on 1/1/80? Well, we know what makes that the case: the ages and temporal distributional properties that things have now makes that the case. What makes it the case that *this* time is present? That question amounts to asking: what makes it the case that things have the properties they have now? Again: their having the fundamental properties that they have makes that the case.

This reverses what is perhaps the usual way of thinking about the moving spotlight. We tend to think, when thinking of the moving spotlight, of things being a certain way at each time, and what is happening *now* amounts to which part of reality is lit up by the spotlight, with the facts about what happened or what will happen being determined by facts about where the spotlight was or will be. I think that is the wrong way to think about it. It is not where the spotlight is, was, and will be that determines what is now, was, and will be the case regarding ordinary matters of fact; rather, how ordinary objects are, were, and will be determines where the spotlight is, was and will be (i.e. determines what times are, were and will be present).

I am not 6ft tall now because the spotlight is shining on a time at which I am 6ft tall. Rather, I am 6ft tall, simpliciter, and that is *what it is* for the spotlight to be shining on a time at which I am 6ft tall. And once we have fixed the properties of every thing, we have fixed the unique time on which the spotlight falls. The spotlight falls where it does, and used to and will fall elsewhere, because of how things are. It is not that how things are is a result of where the spotlight falls, with things having been different because the spotlight used to fall somewhere else; that way of thinking about the moving spotlight should get rejected as soon as we give up on the picture of taking the B-Theorist's metaphysic and simply adding to it the fact that some time is special.

This gives us an answer to the objection that the moving spotlighter has no good account of what presentness *is*. The threat was as follows. The presentist thinks that to be present is just to *be*, while the growing blocker can say that to be present is to be on the edge of being. What can the moving spotlighter say? The threat is that she simply has to take it as a primitive feature of reality that some time is present, with nothing interesting to say about what that amounts to. The threat arises because of the mistaken picture of the moving spotlight view as the B-Theory with an added extra: there seems to be nothing informative to say about what that added extra amounts to, other than that it is a primitive property of presentness. But my moving spotlighter does not take presentness as a primitive feature of anything. There's just the way things are, simpliciter, and something's being present is just a matter of it being the case. (This is what you get from *The Primacy of Present Truth*.) And the way things are simpliciter makes it the case that they used to be and will be different. And for some other time to be such that it was or will be present is just a matter of some other global distribution of properties being a way things were or will be.

4.5 On Endurance

In this section, I want to consider persistence through time and the problem of temporary intrinsics. I will argue that in order to be an endurantist—in order to hold that one and the same thing exists in its entirety from one moment in time to the next, but nevertheless changes over time—one has to be an A-Theorist. Hence that the choice is between being an A-Theorist and being a B-Theoretic perdurantist (or stage theorist[25]). Insofar as one finds the metaphysics of temporal parts unsatisfying, then, this is an argument for being an A-Theorist. Some have argued for an even stronger claim: that in order to be an endurantist one has to be a presentist.[26] But I will argue that the moving spotlight view defended in the previous sections is perfectly compatible with endurantism.

So to review the problem of temporary intrinsics: how can one and the same thing, A, be both F (for A is F at time t1) and not F (for A fails to be F at time t2)? Here are two easy solutions to the problem. The perdurantist says that A itself is never F; rather, the thing that is F is a temporal part of A that exists at t1 but not at t2. Nothing is ever F at one time and not F at another time, for the thing that is F is a mere temporal part of some larger spacetime worm, and it does not exist for longer than it is F. On this metaphysic, I am not 6ft tall. In fact, I am no height. What is 6ft tall is a temporal part of me: the part that is located (only) in the present. Similarly, I was never 5ft tall. What is 5ft tall is a temporal part of me that is located (only) in the past. So the perdurantist simply denies that one and the same thing is both F and not F: something is F and some *other* thing is not F. Another easy solution is presentism. Assume t2 is present. Then the presentist simply says that A is not F. A *was* F, when t1 was present, but is so no longer; and there is no contradiction at all in saying that one and the same thing both is some way and was some other way. On this metaphysic, nothing is ever F at one time and not F at another time, because there always only *is* one time.

The problem of temporary intrinsics really starts to bite when we try to combine non-presentism with endurantism, for then there are many times and one and the same thing has, apparently, incompatible properties at different times. Here are a couple of familiar attempts at reconciling non-presentism with

[25] See Hawley (2001) and Sider (2003) for discussion of perdurantism and stage theory.
[26] See Merricks (1999).

endurantism. They each make no explicit appeal to anything A-Theoretic, and they have not been presented by their defenders as relying on the A-Theory, so I assume they are meant to be neutral on that issue. But I will argue that each of these views combined with the B-Theory is unsatisfying.

Peter van Inwagen argues that apparent intrinsic monadic properties are really relations to times.[27] So when we say that A is F at t1, we are really saying that A bears the *being F at* relation to time t1. Whereas A fails to bear that relation to time t2, which is why we say that A is not F at t2. Sally Haslanger argues that F is as it appears—an intrinsic monadic property—but that nothing instantiates this property simpliciter, but rather it is only ever instantiated in a certain way.[28] So A is F t1-ly, but it is not F t2-ly. Now the fact that A changes is no puzzle, for although A is F at one time and not F at another, this merely amounts to A having F in one way but not another, which is perfectly consistent.

I think that nothing like these approaches is going to work. Granted, each view avoids getting into an outright contradiction, but a satisfactory account of persistence through change has to do more than simply be consistent; it must also account for the *change in appearances*. A theory of how I persist through time must account for not just how it can consistently be the case that I am now 6ft tall but was 5ft tall, but also must account for why, when you look at me now, I appear to be 6ft tall but do not appear to be 5ft tall, and why, when you looked at me when I was ten years, old I appeared to be 5ft tall but did not appear to be 6ft tall. No problem for the presentist: I look 6ft tall now because I *am* 6ft tall now. That is the way I am simpliciter, and my being that way explains why I appear that way. But when I was ten I had different properties; when you looked at me then, I appeared 5ft tall because I *was* 5ft tall—that is how I was, simpliciter, at that time, and my having been that way then explains why I appeared that way then. Similarly, no problem for the perdurantist. Why do I appear to be 6ft tall now? Because the only part of me that is around now to be looked at is 6ft tall, simpliciter. Why did I appear to be 5ft tall when I was ten? Because the only part of me that is around then is 5ft tall, simpliciter.

For the presentist, the change in appearances over time is accounted for by the fact that how things are, simpliciter, changes over time. For the perdurantist, the change in appearances over time is accounted for by the fact that you only ever see a proper part of a persisting thing at any given time, and the different parts you encounter at different times are different ways. But on both van Inwagen's and Haslanger's views, you encounter the entirety of a thing at every time at which you encounter any of it (for they are endurantist views: objects are wholly

[27] van Inwagen (1990). [28] Haslanger (1989).

present at every time at which they exist), and how the objects are, simpliciter, does not change from one time to another. How, then, can a change in appearance be accounted for?

I am 6ft tall but was 5ft tall. So on van Inwagen's view, that is a matter of me bearing the *is 6ft at* relation to the present time t1, and bearing the *is 5ft at* relation to a past time t2. But my standing in those relations to those times is not itself something that is subject to change. (Were it so, the problem of change would simply re-arise: how can one thing both bear this relation to that time and not do so?) I *always* (at least, at every moment at which I exist) bear the *is 6ft at* relation to t1 and the *is 5ft at* relation to t2. *A fortiori* I bear the *is 6ft at* relation to t1 and the *is 5ft at* relation to t2 *now*. Why, then, do I now look 6ft tall and not 5ft tall? I now enter into both relations; why is only one affecting my appearance?

It is tempting to answer: I look 6ft tall now because I bear the *is 6ft at* relation to t1, and it is *now* t1, whereas I do not look 5ft tall because I only bear the *is 5ft at* relation to a time that is not now. But we are considering this account of persistence embedded in a B-Theoretic view, remember. There is no objective now. To say that it is now t1—that the present time is the one I bear the *is 6ft at* relation to—is simply to say that t1 is the time at which I am having *this* thought. t1 and t2 are equally real. The obtaining of the *is 5ft at* relation between me and t2 and the obtaining of the *is 6ft at* relation between me and t1 are both equally real. And when you look at me now you look at the thing that is the other relatum of both those relations. You are looking at *all* of me, and I enter into both those relations to those different and equally real times. There is no metaphysical reason to privilege the obtaining of one relational fact over the other. So why should the appearances reflect only my standing in the *is 6ft at* relation to t1 and not my standing in the *is 5ft at* relation to t2?

Likewise with Haslanger's view. She holds that I have the property of being 6ft tall t1-ly and the property of being 5ft tall t2-ly. But I have those two properties in those two ways now, so why when you look at me now do you only see me being 6ft tall? When t2 was present, I also had those two properties in those two ways, so why when you looked at me then did you only see me being 5ft tall? Why has my appearance changed, when how I am—what properties I have in what ways—has not changed? And again, we are tempted to say: I appear to be 6ft tall when you look at me now because that is the property I have t1-ly, and it is *now* t1. But as before, this is really no explanation. There is nothing special about t1. It is not even the time I am at, for given endurantism I am wholly at both t1 and t2. t1 is real, t2 is real, and there is no more reality to my having *is 6ft tall* t1-ly than there is to my having *is 5ft tall* t2-ly—those are both ways I am, simpliciter, at every time at which I exist. When you look at me now, you look at the whole of me, and the whole of me has *is 6ft tall* one way and the whole of me has *is 5ft tall* another way. And this is true of me whenever you look at me. So why do I appear different from one time to another?

My appeal to temporal distributional properties is inspired by Josh Parsons, who uses them to argue that an eternalist need not be a perdurantist.[29] But Parsons does not combine them with ages and so, as with van Inwagen and Haslanger, his view is that how things are simpliciter is not changing from one time to another. As with their views, then, I do not see how Parsons' view can account for a thing's change in appearance. Parsons attributes to me a distributional property that describes the variation in my height across time. But this is a property that I always have: I am *always* such as to vary across time in height as that distributional property describes. That is the way I am when I am 5ft tall, and it is the way I am when I am 6ft tall. So as above, there is no explanation for why I appear simply to be 6ft tall now and appeared differently when I was ten years old, given that how I am—that I vary with respect to height a certain way across time—is the same at each time on the postulated metaphysic.

Of course, sometimes A appearing different to me on separate occasions can be perfectly explainable even if there is no change in the properties that A has simpliciter from one occasion to another. If I view a stick on one occasion (t1) it can appear straight, while on another occasion (t2) it can appear bent, even though there has been no change in the stick. But that is because the stick is half submerged in water on the second occasion: there has been no change in the stick, but there has been a change in the stick's surroundings, and we have an account (involving the refraction of light) that explains the resulting change in appearances. But this has to be a genuine change in how reality as a whole is if it is to explain the change in appearances in the enduring stick; the B-Theorist cannot account for it. The B-Theorist endurantist, to explain how one and the same thing (the stick) can be both submerged in water (at t2) and not submerged in water (at t1) has to say something like: the stick bears the *being submerged in* relation to the water and t2 but not to the water and t1. But, as above, the stick is *always* such as to bear that relation to the water and t2 but not to the water and t1. Not only is the stick the same way simpliciter at both times, but it relates to other things the same way simpliciter at both times. There has not been any genuine change in how the stick relates to the water, so why the difference in appearances? Similarly, Bob might view A one way at one time and a different way at a later time despite neither A nor its immediate surroundings having changed—but that is because Bob got cataracts, and we have an account of how a subject undergoing

[29] Parsons (2000).

such a change affects how the subject experiences things. But again, if Bob is an enduring entity, this can only explain the change in appearances if Bob has genuinely changed, in the manner the A-Theorist but not the B-Theorist allows. If all that is true is that Bob bears the *has cataracts* relation to the later but not the earlier time (e.g.)—well, Bob always bears that relation to the one time and not the other, so what explains why he views things one way and then another way, given that how he stands to things with respect to the *has cataracts* relation has not changed? Finally, if I view a coin from directly above on one occasion it will appear to me round, whereas if I view it from the side on another occasion it will appear to me elliptical. But in that case there is a difference between the spatial relations that hold between me and the coin on the two occasions, and we have an account of how perception depends on the spatial relations the perceiver stands in to the perceived. And again, if the coin and I are enduring beings then this has to be a genuine change between the relations that hold between me and it, of the kind the A-Theorist but not the B-Theorist will allow. If all that is going on is that the coin and I and time t stand in a certain spatial relation but the coin and I and time t* stand in a different spatial relation, then since those relations always hold between those things, there is again no explanation for the change in appearances. And in general, since the B-Theorist thinks that how a thing A and a thing B are simpliciter—including what relations they stand in to each other and their surroundings—is not something that is subject to change, then I do not see how they could account for the very same thing B appearing differently from the very same thing A from one time to another.

The moral of the story: in order to explain a change in the appearance of some enduring Xs to an enduring subject A, there either has to be a genuine change in the Xs (i.e. a change in how they are *simpliciter*), a genuine change in A, or a genuine change in the relations that hold between them, or between them and their surroundings. The B-Theorist thinks that there are no changes concerning how things are, or how they are related, *simpliciter*, so she can only explain the change in appearances by saying that A encounters different parts of the Xs at different times, and how those parts are (simpliciter) is different, or that it is different parts of A that are doing the perceiving at different times, and how those parts of A are (simpliciter) is different. That is, she has to be a perdurantist. Endurantism requires the A-Theory.

My view, by contrast, faces no problem in accounting for the change in appearances over time. All of me is around just now, and I have a temporal distributional property that describes how my height varies across time, and an age that describes how far through my life I am. Having both those properties entails that I am 6ft tall, not that I bear a certain relation to a time, or that I have a certain property a certain way: that I am 6ft tall, simpliciter. So why do I appear to be 6ft tall, and not 5ft tall, when you look at me now? Because I *am* now 6ft tall, and not 5ft tall, simpliciter. The properties I have, simpliciter, entail that this is how I am, and this accounts for my appearance. Why *did* I look 5ft tall and not 6ft tall when you looked at me when I was ten? Because I had different properties. I had the same temporal distributional property, but I had a different age, and having the combination of those properties entails that the bearer is 5ft tall and not 6ft tall. The properties I had, simpliciter, at that time entailed that I was 5ft and not 6ft tall, simpliciter, at that time, and that is what explains why I appeared the way I did then. So because I undergo a genuine change in properties—because the way I am, simpliciter, is different at the two different times—I have an explanation for the change in appearances.

I have argued that in order to be an endurantist you have to be an A-Theorist: there has to be a genuine change in how things are simpliciter from one time to another to explain how one and the same thing can appear differently from one time to another, and only the A-Theorist can allow for genuine change in how things are simpliciter, for that requires how reality as a whole is to change. Trenton Merricks makes a stronger claim: that only the *presentist* can be an endurantist. After pointing out that the presentist has an easy response to the problem of temporary intrinsics, Merricks says that a non-presentist endurantist avoids the problem by saying that objects have their properties *at times*, and that there is no logical contradiction in an object having a property *at* one time and not *at* another time. But this is not what is distinctive about the non-presentist endurantist, says Merricks, for there is a sense in which *everyone* agrees that objects have properties at times. What is distinctive about non-presentist endurantist metaphysics, says Merricks, is that they *deny* that objects have properties *simpliciter*. He says:

> What is distinctive about the non-presentist reconciliation of change and endurance is not what it affirms—the having of properties at times—but rather what it renounces—the having of properties simpliciter (i.e., not modified in some way by a time). The non-presentist endurantist must deny that an object can have a property simpliciter. [He adds in a footnote: "At least, she must deny this for any 'temporary' property, any property that can be gained or lost."] For if an enduring object O could have a property F simpliciter and lack it later, and if presentism were false, then we'd have the contradiction that O is both F simpliciter and it is not the case that O is F simpliciter. Obviously, this is contradictory even if O has some other properties, such as being F at t.[30]

[30] Merricks (1999, pp421–38).

He then proceeds to argue that non-presentist endurantism is ultimately incoherent because as well as renouncing the having of properties simpliciter, the non-presentist endurantist, for similar reasons, has to renounce the having of parts simpliciter. But endurance, Merricks argues, is best characterized as the doctrine that things persist and that, at all times at which they exist, all the parts they have *simpliciter* exist at that time. In renouncing the having of parts simpliciter, says Merricks, one renounces the claim that objects endure, which presupposes that objects have parts simpliciter. Thus non-presentist endurantism renounces endurantism, rendering it incoherent.

I think Merricks is mistaken in thinking that the non-presentist endurantist has to deny that a thing can have properties (or parts) simpliciter. My moving spotlighter accepts that things have properties simpliciter, not merely at a time. Indeed, the properties a thing has simpliciter are the ones it has *at* the present time. That is why she accepts *The Primacy of Present Instantiation*. My moving spotlighter thinks exactly what the presentist thinks: that objects have properties simpliciter, and that the way things are simpliciter is the way they are now, but that objects *change* with respect to what properties they have simpliciter. The difference between the presentist and my moving spotlighter does not concern whether they allow instantiation of properties simpliciter, but rather concerns what things they think have properties simpliciter: for the presentist, only present things have properties simpliciter, whereas for my moving spotlighter both present and non-present things have properties simpliciter.

Merricks says that the non-presentist endurantist has to deny that objects have properties simpliciter because "if an enduring object O could have a property F simpliciter and lack it later, and if presentism were false, then we would have the contradiction that O is both F simpliciter and it is not the case that O is F simpliciter."[31] But I deny this. If an enduring object O is F now but was not F, then it is F simpliciter. But from the fact that O *was* not F we cannot conclude that it is not the case that O is F simpliciter. Instantiation is present instantiation. How things are simpliciter is how they are now. So O is F, simpliciter. And it *was* the case that O is not F, simpliciter. But O's failing to be F is merely a way O was, and this is now nowhere to be found in reality, not even in the past.

The rejection of *Past Record* is what is enabling the moving spotlight view to avoid this problem with change over time, just as it enabled it to avoid McTaggart's paradox. (It should come as no surprise that the two problems get the same kind of solution, given the affinity between them that was noted in chapter two.) Merricks is assuming that the moving spotlighter, in admitting the reality of the

[31] Merricks (1999, pp421–38).

past, has to admit the reality of O's failing to be F, since that is something that was the case. If O's failing to be F was a matter of O's not being F simpliciter, this means admitting the reality of O's not being F simpliciter, which gets us into contradiction. But this assumes that the moving spotlighter accepts *Past Record*: that she holds that because something was the case, it *is* the case in the past. If the non-presentist endurantist accepts *Past Record* then I agree that she is vulnerable to Merricks' argument. But my moving spotlighter rejects that principle and hence is not so vulnerable. She thinks that there is no reality to O's not being F. Accepting the past, for my moving spotlighter, amounts to accepting the existence of past entities. But she does not accept that how things were remains a part of how reality as a whole is. How things are, simpliciter—whether those things be located in the past, present, or future—is how they are *now*. O is F—end of story. O *was* not F, and the *fact* that O was not F is made true by the way O now is. But O's not being F is merely a way reality was, and is now no part of reality. Thus my moving spotlighter avoids the puzzle of change exactly as the presentist does: the only way things are in reality is the way they are now. They used to be a different way; but there is no contradiction in some thing's being a certain way and it *having been* a different way.

To sum up the moral of this section: if one and the same thing appears different from one time to another to one and the same observer, there must be a change over time in how that thing is simpliciter, or in how the thing doing the observing is simpliciter, in order to account for that change in appearances. The perdurantist accepts this conditional and rejects the antecedent: it is not the very same thing that appears different from one time to another or that is doing the observing from one time to another. The presentist and my moving spotlighter accept the conditional and accept both antecedent and consequent: as time passes, how things are simpliciter changes. Endurantist B-Theorists must reject the conditional, because they are committed to the antecedent (which follows from endurantism with the obvious truth that things appear to change), but they cannot accept the consequent, since according to the B-Theory how things are simpliciter is not subject to change. This leaves these views unable to account for the changes in how things appear over time, which is unacceptable. Thus, if you are going to be an endurantist, you ought to be an A-Theorist. And, in particular, you ought to be an A-Theorist who accepts that things change with respect to how they are simpliciter, as both the presentist and my moving spotlighter do.[32]

[32] Incidentally, if the arguments in this section are correct then I think they rule out the possibility of qualitatively heterogeneous extended simples, since there is no spatial equivalent of

4.6 On History's not Changing

In this section I want to tackle an objection to my use of temporal distributional properties (TDPs) that was put forward by Jonathan Tallant and David Ingram.[33] The problem, on inspection, turns out to be a problem for all A-Theories. I think the problem is solvable, and that solving it brings to the fore some interesting issues.

I said above that things do not change with respect to what temporal distributional properties they have. This had better be true, because otherwise we would get a violation of an overwhelmingly plausible principle of tense logic: that if something is now the case, then it always will be true that it was the case. That is, if p is now true, then for any future time t, "p was true" will be true when t becomes present.

For suppose that something changed which TDP it instantiated. So suppose I currently have a TDP that describes my height as growing from 2ft to 6ft over a certain period of time, but that I am about to lose that property and have it replaced with a TDP that describes my height as growing from 2ft to 5ft and then halting there. In that case, while I am now 6ft tall, it will be the case, once my TDP

the genuine change across time that the A-Theorist believes in. So consider Parsons' (2004) view that qualitatively heterogeneous extended simples instantiate a distributional property that describes their variation across space. I say that this cannot explain the heterogeneity, because the *whole* of the thing in question instantiates that property simpliciter—so how can we explain why it appears red here but blue there (say)? Similarly, McDaniel (2009) following a proposal by Ehring (1997), says that if there is an extended simple that is red here and blue there, that is because it has a particular redness trope that is located here and a particular blueness trope that is located there. But again, the *whole* of the thing (since there is only one part) has both the redness trope and the blueness trope; so what explains why it is not (impossibly) both wholly red and wholly blue? How does the tropes' being *located* where they are explain anything about why the bearer of those tropes is as it is at those locations? Is it not simply having a property that explains why a thing is a certain way, not where the property happens to be? (Plato and Aristotle disagree over where properties like blueness and redness are, but they are not disagreeing over where things appear to be blue and red.) In that case, since the whole of the object has the redness trope and the whole of the object has the blueness trope, I cannot see how there is any explanation for why it appears to be red here but not there, and blue there but not here. Things extended over space cannot admit of genuine variation across space in the same way that enduring things can (so long as the A-Theory is true) admit of genuine variation across time. The change in appearances of enduring things over time can only be explained by a genuine change in what properties they instantiate simpliciter from one time to another. There is no spatial analog of that, and so I think qualitatively heterogeneous extended simples are impossible.

[33] Tallant and Ingram (2012). They put forward their objection to my use of these properties as employed by the presentist in Cameron (2011), but I think their objection also targets the moving spotlight version of the view. (Other objections to the use of temporal distributional properties, such as that advanced in Tallant (2012), clearly only apply to the presentist's use of them. Insofar as those objections are successful (an issue I will take no stance on here), this gives us another reason to be a moving spotlighter rather than a presentist.)

has changed, that I have never been and never will be 6ft tall, since the new TDP has my height vary only between 2ft and 5ft over time. In that case, something that is true now will come to never have been. That is absurd: what is true now will, no matter how else the future unfolds, always remain a part of the past. If things can change their TDPs then this principle of temporal logic can be violated. So things must not be able to so change. But is that plausible, given that these are *accidental* properties? As Tallant and Ingram point out: I am committed to thinking that *some* properties of things change, because everything's age changes all the time. So why can a thing's TDP not change?

Before I offer my defense, I will argue that the problem Tallant and Ingram are raising is by no means a problem only my view faces. *Any* A-Theory, in allowing that how reality is as a whole can change, faces a version of this puzzle.

Suppose first that we are presentists and truthmaker theorists. In that case, we think that present reality—which is reality simpliciter—contains entities which ground the historical truth that, for example, Caesar crossed the Rubicon. There is disagreement as to what kind of thing does the grounding, but if we are truthmaker theorists then *some* such thing does. But if we are presentists then we also think that what there is—by which I mean, what there is *simpliciter*—changes: being is *present* being, and things come into and go out of being. Caesar existed, but he exists no more; and it is not just that he presently existed but does not exist presently any more (everyone agrees with that!)—it is that he existed, simpliciter, but now does not exist, simpliciter (and not everyone agrees with that). But if what there is simpliciter can so change, then what is to stop the presently existing truthmaker for <Caesar crossed the Rubicon> going out of existence and being replaced by an entity that makes it the case that Caesar never crossed the Rubicon? But if that will happen, then something that is now a part of history will not in the future be a part of history: and so we would have a violation of the principle cited above—that if something is now the case, it always will be that it was the case.

Suppose now we are the presentist who eschews truthmakers and thinks that there are simply brute tensed facts. So we say that it is true that Caesar crossed the Rubicon, and this is simply a brute fact about how reality is, with no further ground. Still, what is true changes, according to the presentist, so why not this? While it is now the case that <Caesar crossed the Rubicon> is brutely true, what ensures that tomorrow it will not be brutely true that Caesar never crossed the Rubicon? But were this so, our principle of tense logic would again fail.

And in fact *any* A-Theorist, I think, faces this challenge, for every A-Theorist thinks that how reality is as a whole is subject to change. Why then cannot reality change with respect to whatever features the truth of historical truths are sensitive to? The moving spotlighter thinks that "There were dinosaurs" is true because there are dinosaurs in the past. But since I know that reality will be different in at least *some* respects tomorrow (if nothing else, a different time will be objectively present), how can I guarantee that it will not change in this respect: perhaps tomorrow there will no longer be dinosaurs anywhere in reality, in which case something that is true now will never have been true tomorrow.

The moral of the story: *all* A-Theorists must rule out certain changes from occurring if they are to ensure the truth of plausible principles of tense logic. If you believe in brute tensed facts, you must place constraints on how those tensed facts can change; if you believe in truthmakers for tensed truths, you must place constraints on what changes can occur regarding what truthmakers exist. My ban on things changing their temporal distributional properties is simply an instance of what every A-Theorist must do.

Thankfully, I think that it is no cost to deny that such changes can occur, and that the argument to the contrary illegitimately attempts to invoke a kind of tensed claim that goes beyond what is sanctioned by the brute tensed facts, or the extant truthmakers for tensed truths (whichever you appeal to to make true truths about how things were or will be).

Consider again the brute-truth presentist.[34] She thinks that reality looks something like this: there are cars and cats, but no dinosaurs or lunar colonies. But it is brutely true that there were dinosaurs and will be lunar colonies. But the brute tensed facts that partly make up present reality (which is reality simpliciter) do not end with these relatively mundane past and future facts concerning dinosaurs and lunar colonies: they will also include facts about what *were* and *will be* the past and future truths. Just as (present) reality includes the brute tensed fact that there were dinosaurs, so does it include the brute tensed fact that it always will be the case that there were dinosaurs. So what stops the brute tensed fact that there were dinosaurs changing and it becoming the case that there never were any dinosaurs? The presently obtaining brute tensed facts stop that, for one of them says precisely that this change will not happen.

Similarly for any A-Theorist, presentist or otherwise, who posits truthmakers for historical truths. She thinks that reality contains things that make it true that there were dinosaurs and things that make it true that there will be lunar colonies. But the extant truthmakers for tensed truths do not end there: there is also a thing that makes it true that there will always be a thing that makes it true that there were dinosaurs. So what stops the truthmaker for <There were dinosaurs> going out of existence and a truthmaker for <There were never dinosaurs> coming into

[34] Such as Merricks (2009, ch.6).

existence? The thing that makes it true that there will always be a thing that makes it true that there were dinosaurs stops that.

We are tempted to look at an ontology of truthmakers for historical truths and ask: but what if there is a change in what truthmakers exist? Well, what *is* change? Change is a matter of something being true at one time and not at another, and it is these very truthmakers that we are asking about that account for such facts. So if there is to be a change in which of these truthmakers exist, it is because one of them *makes* it true that there will be such a change. That is the only place change can come from, because *all* tensed truths come from the existence of these truthmakers. So to see how the existence of the truthmakers will change, we need only look to those very truthmakers. And so to rule out violation of the principle of tense logic—to guarantee that if something is the case, it always will be that it was the case—the A-Theorist merely has to be careful when she specifies her ontology. If she postulates something that makes it true that it was the case that p, she had better be careful to also posit something that makes it true that it will always be the case that there is something that makes it the case that it was the case that p. So long as she is so careful, there can be no violation of the principle that appeared under threat. And likewise, *mutatis mutandis*, for what brute tensed facts the non-truthmaker A-Theorist postulates in reality.

I see nothing methodologically suspect about the A-Theorist simply postulating that this is how things are, in order to uphold the relevant principle of tense logic. *Of course* she should let her beliefs about what kinds of change are possible guide her theory about what tensed truthmakers[35] exist (or what brute tensed facts obtain), given that it is these tensed truthmakers (or brute tensed facts) that account for change, on her metaphysic. And whatever strength she thought attached to the principle of tense logic, she should think the same strength attaches to the above principles concerning tensed truthmakers or brute tensed facts. If all she is concerned with is the truth of the material conditional that if something is true now then it always will be the case that it was true, then her theory merely needs to entail the truth of material conditionals such as: if there is a truthmaker for "Caesar crossed the Rubicon" then there is a truthmaker for "There always will be a truthmaker for 'Caesar crossed the Rubicon'." If, on the other hand, you think the principle of tense logic holds with metaphysical necessity, then your theory should claim that this connection between what truthmakers there are is a metaphysically necessary one. If you think there is

[35] By "tensed truthmaker" I mean simply a truthmaker for a tensed truth.

some special status of "being a theorem of tense logic" that the principle has, then you should think the link between truthmakers has this special status also.[36]

It is tempting to see a threat to the principle of tense logic because it is tempting to think that there can be changes that are not grounded by the brute tensed facts that obtain or the extant tensed truthmakers. We see this truthmaker for "Caesar crossed the Rubicon," and we worry about what might happen if it fails to be around. We are told not to worry, because here is another truthmaker that says that the first one will always be around, thus securing the truth of "It will always be the case that Caesar crossed the Rubicon." But intuitively, that does not address the worry. For what if *neither* of these truthmakers is around in the future? What if *all* the truthmakers for tensed truths disappear? And it is no good, so the thought goes, to be told that there is *now* something that makes it true that this will not happen, for what happens if that goes out of existence too?

This is a seductive line of thought, but it simply has to be abandoned. It attempts to invoke tensed claims that are not grounded in the existing truthmakers for tensed truths (or the obtaining brute tensed facts), and the A-Theorist should simply deny that this makes any sense. The extant tensed truthmakers are the sole source of tensed truth, she claims (or the obtaining brute tensed facts), and these tell you everything there is to know about what was or will be the case, *including* the facts about what tensed truthmakers did or will exist (or what brute tensed facts did or will obtain). Any attempt to ask a question about what was or will be the case that is not to be settled by what tensed truthmakers exist (or what brute tensed facts obtain) is simply illegitimate by the A-Theorist's lights. And so, as long as she is careful, in the sense just spelled out, about what tensed truthmakers (or what brute tensed facts) she posits, then there is simply no threat here.

[36] This may be where the strongest objection can be made. Suppose you held that tense logic is really a logic, and not just a description of certain facts about how time in fact is, and that it is a theorem of this logic that if p is now true then it always will be the case that p was true. My suggestion is that you then also take it to be a theorem of tense logic that if there is a truthmaker for p, there is also a truthmaker for "It always will be the case that there's a truthmaker for 'It was the case that p'." But, you might object, there cannot be *theorems* concerning *what exists*: ontological claims like this are never a matter of *logic*. On these grounds, an objector might see a problem that arises particularly for *truthmaker* versions of the A-Theory. I am unsympathetic to this objection. Even if our principle has a special kind of status of theoremhood that is more than mere truth (and more than mere metaphysical necessity, even), I don't see a good reason for denying that the link between tensed truthmakers can have the same status. Why can there not be theorems of tense logic that concern what exists, when what exists includes tensed truthmakers? One might hold that in ordinary (non-tense) logic there are no theorems concerning what there is. But if so, that is because ordinary logic is *topic neutral*. But of course tense logic, if there is such a thing, is *not* topic neutral: it is *about* time. So just as it can include theorems about time, I do not see why it cannot also include theorems about tensed truthmakers.

So to return to my preferred metaphysic. I say I have a TDP and an age: together, they make it true that I am over 30 years old and 6ft tall, and that I was 10 years old and 5ft tall. Tallant and Ingram ask: how can I rule out my changing my TDP, and coming to instantiate one that makes it true that I was never 5ft tall? Answer: my having the TDP I presently have makes it true that I will always have it.

My present age makes it true what ages I had and will have. When I had the property *being 30 years old*, my having that property made it true that I will have the property *being 33 years old* three years from then. And just as the age I have now grounds the facts about what ages I had or will have, so does the TDP I have now ground the facts about what TDPs I had or will have. And in particular, it grounds the fact that I always had and always will have the one I presently have. You can't vary in what TDP you have from one time to another, because that would be to vary across time in how your intrinsic nature is across time. Of course that cannot happen, because once you have settled how your intrinsic nature varies across time, that also settles how at other times your intrinsic nature varies across time.

Compare the spatial case. Consider a really long bar that stretched the length of Scotland, and which is black and white striped, with stripes thick enough to be the length of cities. Looking at the bar in Glasgow, I can say: (i) the bar is black here, (ii) the bar is white in Edinburgh, and (iii) the bar is black and white striped across space. Statement (iii) describes how the bar varies across space *here* in Glasgow, but it also truly describes how the bar varies across space in Edinburgh, and indeed anywhere. When we are talking about how the bar is *across* space, it does not matter *where* we are assessing things from. When claiming that it is black here, or white here, my claim is sensitive to where "here" is, since how the bar is with respect to color varies across space; but when claiming that it is such-and-such a way across space, the location of assessment just drops out. How things are *across* space is not the kind of thing that can vary from one place to another.

Likewise, how things are across time is not the kind of thing that can vary from one time to another: settle how a thing is across time and you thereby settle how it is across time *at other times*. Which is just to say that having a particular TDP at a time makes it true that you always had, and always will have, that very TDP. And that ensures that our principle of tense logic will never be violated.

One more thing must be done before the problem can be considered adequately addressed. I said above that the appearance that the principle of tense logic might be violated was due to a seductive, but illegitimate, thought: that there could be changes concerning what truthmakers for tensed truths exist (or changes concerning what brute tensed facts obtain) that are not themselves simply a result of what truthmakers for tensed truths there are (or what brute tensed facts obtain). That thought *is* illegitimate, but it is *so* seductive that I think we cannot rest until we have accounted for it. This is what I will attempt to do now.

I aim to borrow a methodological trick from Kripke: that when you are denying the possibility of a scenario that really seems possible, you should identify a genuinely possible scenario that you are plausibly confusing with the scenario deemed impossible. So, for example, Kripke tells you that it is impossible that water is not H_2O; but he accounts for intuitions to the contrary by claiming that you are confusing this scenario with the genuinely possible scenario that something has all the phenomenal surface qualities of water but is not H_2O.[37]

I deny that it is possible that the truthmaker for "Caesar crossed the Rubicon" go out of existence and be replaced by a truthmaker for "Caesar never crossed the Rubicon." And yet I concede that it genuinely seems possible. But don't worry: there is a genuinely possible scenario that you are confusing with this impossible one. The genuinely possible scenario is that there is now a truthmaker for "Caesar crossed the Rubicon" but that there hyperwill be no truthmaker for "Caesar crossed the Rubicon" and instead a truthmaker for "Caesar never crossed the Rubicon."

Here I am invoking the notion of hypertime, which is what one needs to make sense of the past changing.[38] While ordinary tenses let us describe our past and future, hypertenses let us describe a kind of change to history as a whole. So while it will always be the case that Caesar crossed the Rubicon, this is compatible with it being true that it hyperwill be the case that Caesar never crossed the Rubicon. Likewise, while what tensed truthmakers there presently are determines what tensed truthmakers there were and will be (since those tensed truthmakers determine *all* the facts concerning what was and will be), there now being these truthmakers does *not* determine what tensed truthmakers there hyperwere or hyperwill be. So there is a truthmaker for "Caesar crossed the Rubicon," and there always will be such a truthmaker: but maybe it hyperwill be the case that there is no such truthmaker.

I am not claiming that there is such a thing as hypertime. But I think it is metaphysically possible, and that is all I need. While it *seems* possible that what tensed truthmakers there are can change in a way not governed by those tensed truthmakers, what is *really* possible is merely that there are such tensed truthmakers, and that history is (thereby) thus-and-so, but that there hyperwill be different tensed truthmakers, hence that it hyperwill be the case that history is different from how it now is.

[37] Kripke (1980, esp. Lecture 3).
[38] See van Inwagen (2010) and Hudson and Wasserman (2010).

So on my own preferred metaphysic: it might be that I hyperwill have a different TDP, but that's no problem, for it leads to no violation of any plausible principle of tense logic. Changing my TDP would cause trouble; but I *will not* change my TDP, and I know that this is the case because my having my TDP *makes* this the case. So we have both an explanation for why Tallant and Ingram's problematic scenario cannot arise, and also an explanation for the seductiveness of the intuition that it may.

4.7 Meeting the Desiderata

Let me end by returning to the desiderata from chapters two and three that we said any successful moving spotlight metaphysic had to meet, to say why I think the metaphysic put forward in this chapter meets them.

In chapter two we said that the moving spotlight had better be able to distinguish itself from the stuck spotlight with a fancy semantics. The stuck spotlight view says that there is a privileged present, but that how reality as a whole is does not change; *a fortiori*, which time is objectively privileged does not change. However, the stuck spotlight view allows you to correctly *say* that other times were or will be present, for it offers as the truth-conditions for the sentence "t was/will be present" that t be earlier/later than the time that is (always) objectively present. The worry is that this stuck spotlighter will be able to *say* all the things the moving spotlighter wants to say. So how can the moving spotlighter justify the claim that her metaphysic is different: that there is genuine *passage*?

I think my moving spotlighter adequately distinguishes herself from the stuck spotlight theorist precisely because she is *not* simply taking the B-Theorist's metaphysic and *adding* to that metaphysic that one time out of many be objectively privileged. For the stuck spotlighter, to say that another time will be present is just a way of speaking, not reflective of anything in the metaphysics. Nothing would be lost—you would not do a worse job of describing reality—if you did not adopt the fancy semantics and simply gave a tenseless description of reality, saying what there is, and what things are like, including that one time is (tenselessly) present. By contrast, in saying that things *were* a certain way, my moving spotlighter is saying something about the very nature of things. It is of the essence of things that have such-and-such a temporal distributional property and so-and-so an age that they are a certain way now, but it is also of the essence of things that have those properties that they were and will be some other way. Since my moving spotlighter thinks that things have properties simpliciter, she *can* give a tenseless description of reality: she can say what exists (simpliciter), what those things are like (simpliciter), and what time is present (simpliciter). But to merely give this tenseless description of reality is to miss out something crucial about reality: it is to leave out important information about the nature of things—that the nature of things fixes that they used to, and will be, other than they are. That is why the moving spotlight is genuinely different from the stuck spotlight, and why it is a view on which there is genuine passage: the fact that things were and will be different is a consequence of the very nature of things, not something that we can simply correctly say as a result of a clever semantics.

Let us look now at the four worries for the moving spotlight that were initially mentioned at the end of section 3.1. The first worry was that the moving spotlighter, in order to say that the spotlight used to be elsewhere, has to invoke some primitive tense ideology—and that this is then the only work that notion does. That both seems not worth paying the ideological price for and—this is the second worry—it threatens to render the extra ontology of the moving spotlighter redundant; for once we have primitive tense, why not simply say that there are primitive tensed facts about what used to and will exist?

These two worries, I think, relied on thinking of things in the Quine–Lewis–Sider model, and are no threat to the view put forward here. My moving spotlighter, in the light of chapter 3, rejects thinking of things in terms of ideology versus ontology, and instead thinks in terms of what the fundamental facts are. And the fundamental facts, she thinks, concern solely what there is: certain objects (such as dinosaurs, persons, lunar colonies, and the spatio-temporal regions in which they are located), and certain states of affairs (such as the states of affairs of those objects being located at those regions, and of having the ages and temporal distributional properties that they have). There are no fundamental tensed facts: nothing of the form "such-and-such was/will be the case" is fundamental; all such facts are made true by things having a certain age and distributional property and their being located where they are. So there is no costly ideological notion introduced to do a minimal amount of work. *Ages* are introduced, of course: they are postulated as constituents of some of the states of affairs whose existence is taken to be fundamental. But ages do *lots* of work: they play a role in making true every truth about how things were or used to be, not merely in making it the case that the spotlight used to be elsewhere. It should also be clear from section 4.4 why the extra ontology of the moving spotlighter is not redundant. Past and future things are a certain way now, and their being so plays a role in making true claims about how things were and will be. Again, the threat that non-present entities are redundant results from thinking of things in terms of the Quine–Lewis–Sider framework. Sider said that "the spotlight theorist can accept the [B-Theorist's] reduction of tense for all tensed statements except those concerning presentness."[39] That is to think of the spotlighter as accepting *Past Record* for most claims, but dealing with claims about the moving of the spotlight itself by invoking this bit of primitive tensed ideology. My moving spotlighter is doing nothing like this: she does not accept the B-Theorist's reduction of tense for even ordinary claims like "Ross used to be 5ft tall," for she thinks it is simply false that I am 5ft tall before the present time. I am not 5ft tall anywhere in reality. That is merely a way I was. What *makes* it true that this is a way I was is the way I am now. Claims both about how ordinary concrete enduring things used to be and why other times used to be present get treated in exactly the same way: in neither case does the moving spotlighter accept the B-Theoretic reduction that *what it is* for things to have been that way is for them to be that way at some time before the present, and in each case she accepts that what *makes it the case* that things were that way is things now having the temporal distributional properties and ages that they have. And accepting non-present entities plays an important role in that truthmaking account.

The other two worries from the end of section 3.1 concerned how the moving spotlighter treats change. The threat was that changes in ordinary things get treated as the B-Theorist treats them: simply as variation in how those things are across time. The genuine tense in reality concerns change not in the intrinsic natures of ordinary things, but rather change in what time is present. This both looks ad hoc—why treat the two types of change differently?—and it makes the genuine tense in reality removed from our initial concerns when theorizing about time: namely, change in the ordinary goings on around us. It should be clear now that these worries do not arise for the version of the moving spotlight defended here. As I said above, changes in ordinary things are not treated in a different manner from changes in what time is present. And there *is* genuine change in the properties of ordinary things. How things like you and I are, simpliciter, changes. I am 6ft tall, simpliciter. I didn't used to be. Once again, the threat arose from thinking of the moving spotlight view as B-Theoretic eternalism with an extra component, where my being 6ft tall and my being 5ft tall are both real, and where which is the case *now* depends on where the spotlight falls. That way of thinking about the moving spotlight should be rejected. According to the moving spotlight theory defended here, my being 5ft tall is simply no part of reality. I am 6ft tall, simpliciter, and that is the only way I am with respect to height. But I used to be different—so there is genuine change exactly where we need it to be, in the ordinary matters of fact whose apparently changing status made us theorize about time in the first place.

[39] Sider (2013a, p259).

4.8 Summary

Here, for summary, are the main theses of the moving spotlight view defended in this chapter, grouped by subject matter.

4.8.1 Theses Concerning Truth and Existence

- The fundamental notions of truth and falsity are truth simpliciter and falsity simpliciter. *Truth at a time* is a derivative notion: p is *true at* time t iff (i) t is the present, and p is true simpliciter, or (ii) t is a past/future time, and p was/will be true simpliciter when t was/will be present.
- There is a way reality is, simpliciter. Reality's being some way *at a time* is a derivative notion: reality is a certain way *at time* t iff (i) t is the present, and reality is that way simpliciter, or (ii) t is a past/future time, and reality was/will be that way simpliciter when t was/will be present.
- Something is true, simpliciter, iff it corresponds to how reality is, simpliciter. Something was/will be true simpliciter at time t iff it corresponds to how reality was/will be when t was/will be present.
- The fundamental notion of instantiation is instantiation simpliciter. A thing's being some way *at a time* is a derivative notion: x is F *at time* t iff (i) t is the present, and x is F simpliciter, or (ii) t is a past/future time, and x was/will be F simpliciter when t was/will be present.
- There is such a thing as existence simpliciter.

4.8.2 Theses Concerning Temporal Ontology

- Amongst the things that exist simpliciter are non-present entities: that is, things whose location in spacetime does not overlap the present.
- What substances exist simpliciter does not change. If "A exists" is true simpliciter and "A" refers to a substance, then it has always been the case and always will be the case that "A exists" is true simpliciter.
- What states of affairs exist simpliciter does change, as how substances are simpliciter changes. If A is F simpliciter then there exists simpliciter a state of affairs that makes it true that A is F. If A has not always been F then this state of affairs has not always existed.

4.8.3 Theses Concerning Properties and Change

- Everything that exists simpliciter is some way simpliciter.
- Every substance that exists simpliciter instantiates simpliciter an age.

- Every substance that exists simpliciter instantiates simpliciter a temporal distributional property.
- Substances do not change with respect to their temporal distributional property: if A instantiates simpliciter a certain temporal distributional property then A always has instantiated simpliciter, and always will instantiate simpliciter, that temporal distributional property.
- Substances change with respect to their ages. In fact, such change is constant: every substance has a different age property at every moment of its existence. If A has a certain age simpliciter, A has never previously had simpliciter, and never again will have simpliciter, that particular age.
- Substances stand, simpliciter, in a location relation to a certain region of spacetime. Things do not change with respect to their spatio-temporal location: if A is located simpliciter in spatio-temporal region R, A always has been, and always will be, located simpliciter in R.

4.8.4 Theses Concerning Truthmaking and Determination

- For every substance, the intrinsic nature of that substance is determined by its age and temporal distributional property. That is, if "A is F" is true simpliciter, and "F" is an intrinsic predicate, then "A is F" is made true by A having simpliciter the age and temporal distributional property that it has.
- The way substances were and will be intrinsically is determined by the ages and temporal distributional properties they had and will have. If "A was F" or "A will be F" is true simpliciter, and "F" is an intrinsic predicate, then "A was F" or "A will be F" is made true by whatever makes it true that A had or will have simpliciter the particular age and temporal distributional property it did or will have.
- It is of the nature of temporal distributional properties that their bearers do not change with respect to them; that is, temporal distributional properties are essentially such that things do not change with respect to them. Hence, that a substance S does not change with respect to what temporal distributional property it has simpliciter is made true by S's now instantiating the particular temporal distributional property it instantiates. That is, if A instantiates simpliciter a temporal distributional property F, then the state of affairs of A being F makes it true that A always has instantiated simpliciter, and always will instantiate simpliciter, F.
- It is of the nature of location relations that their relata do not change with respect to them. Hence, that a substance S does not change with respect to what region of spacetime it is located in simpliciter is made true by S's now

standing simpliciter in the location relation to the particular region of spacetime in which it is located. That is, if "A is located in spacetime region R" is true simpliciter, then the state of affairs of A bearing the location relation to R makes it true that "A is located in spacetime region R" always has been, and always will be, true simpliciter.

- It is of the nature of ages that their bearers change with respect to them in a predictable manner. That is, for any substance S, and any age property A1, and any duration of time n years, there will be some age properties A2 and A3 such that it is metaphysically necessary that if S instantiates simpliciter A1 then S instantiated simpliciter A2 n years ago, and S will instantiate simpliciter A3 n years from now. Hence facts of the form "S did/will have age A" are made true by the state of affairs of S's having simpliciter the age it has.
- There is no fundamental property of presentness. Facts of the form "Time t was/is/will be present" are all derivative, and true in virtue of the fact that the way reality is at time t is a way reality was/is/will be. In particular, for every time t there is a list of propositions of the form "A is F," where A is an ordinary substance (not a time) and F an ordinary property (not an A-property such as *being past*, or *being present*, or *being future*), such that every proposition on that list is true at t, and there is no time other than t at which all those propositions are true. There is nothing more to t's having been present, or to t's being present, or to the fact that t will be present, than that the propositions on that list were true simpliciter, or are true simpliciter, or will be true simpliciter. Hence, facts of the form "Time t was/is/will be present" are made true by substances having the temporal distributional properties and ages they have, since that is what makes true every fact about what ordinary properties are had by ordinary substances at each time.

5

The Open Future

In this chapter I turn my attention to the distinction between the fixed past and open future, and argue that the moving spotlighter can give as good or better an account of this distinction than her rivals. After a brief introduction in §5.1, I first consider in §5.2 accounts whereby reality literally branches, containing multiple real future histories. I argue that each such account faces problems and moreover that none of them can actually account for all the ways in which the future might be open. Following a view I have defended with Elizabeth Barnes, I take the openness of the future to consist in it being metaphysically unsettled what will happen. But in §5.3 I argue that there are two ways in which reality may fail to settle some fact: it may leave it *indeterminate* whether that fact obtains, or there may be *no fact of the matter* whether that fact obtains. I illustrate the difference by pointing to various cases in metaphysics. This means that there are two ways in which the future might be open: future contingent claims might be indeterminate, or there might be no fact of the matter concerning future contingents. In §5.4 I argue that the indeterminacy view sits better with the moving spotlight view while there being no fact of the matter sits better with the growing block view. So much the worse for the growing block, I will argue in §5.5, as the indeterminacy view does a better job of accounting for the phenomenon of the open future. Thus what seemed like an advantage of the growing block over the moving spotlight—that it can account for the metaphysical asymmetry between the fixed past and open future by positing an ontological asymmetry between extant past ontology and non-existent future ontology—actually turns out to be an advantage for the moving spotlight theory. I conclude the section by showing how to adopt the indeterminacy account of openness given the particular metaphysic defended in chapter four, and why doing so is still consistent with the truthmaking project of chapter three.

5.1 Openness and Ontology

In chapter four a metaphysic was proposed whereby how things are now (which is how things are simpliciter) makes true all the truths about how they were *and will be*. But some A-Theorists might object that the second half of this is work that should not be done: that while of course reality must account for what *was* the case, we ought *not* to demand that it is rich enough to settle truths about what *will* be the case, because there simply are no such truths to be settled—the future is open, in contrast to the fixed past.

This line of thought is what leads some A-Theorists towards the growing block metaphysic: that the past and present are real, but that there is no future. There are past entities like dinosaurs, and present entities like you and me, but there are no future entities like lunar colonies. The thought is that there is a metaphysical difference between claims about the past and present on the one hand and claims about the future on the other hand—the former are fixed whereas the latter are open—and this metaphysical difference must be underwritten by an ontological difference: past and present entities are real, and hence there to serve as truth-makers for the true claims about what is happening now or what did happen in the past, whereas there are no future entities to render true any claim about how things will be, thereby leaving it open whether things will be that way or not. Thus we have growing blockers like Joseph Diekemper arguing that any view on which there is an ontological symmetry between past and future—as with both presentism and the moving spotlight—cannot account for the metaphysical asymmetry between the past and future.[1]

In this chapter I will argue against this. The metaphysical asymmetry—that the past is fixed and the future open—can be secured despite thinking that past and future entities are equally (un)real. In particular, I will offer an account of the open future that is perfectly compatible with the moving spotlight metaphysic defended in the previous chapter. I will also argue that there are reasons to prefer this kind of account of the open future to that which naturally sits with the growing block theory.

5.2 Against Branching

One view which aims to reconcile the reality of the future with openness in what will happen is a branching futures ontology. On such a view, *every* possible future exists, and openness consists in there being multiple futures. This is not the kind of view I will defend, so in this section I will argue against such views.

There are different ways in which a branching time metaphysic can be developed; some objections I raise here will apply to any branching time model, but others will be specific to particular ways of developing the idea.

[1] Diekemper (2005).

I will not aim at exhaustiveness here, but I think I am covering the most plausible versions of branching in what follows.

Branching time need not require a tensed reality. One can have a B-Theoretic branching time metaphysic, according to which time has a tree-like structure that never changes. What branches there (atemporally) are is the same from one moment to the next; furthermore, what node on the tree is *present* is not a matter of how the world objectively is, but is simply a matter of our perspective on the world. If I claim that a particular node on the tree, N, is present, then the truth-conditions of my utterance are simply that the event of my uttering that is located at N. This view is very much like standard B-Theoretic eternalism; the only difference is that *being temporally distant from* is not an equivalence relation—events e1 and e2 might both be after the present, in virtue of being further up the tree, but e1 is not before, after, or simultaneous with e2, in virtue of their being located on different branches.

So suppose reality looks like this:

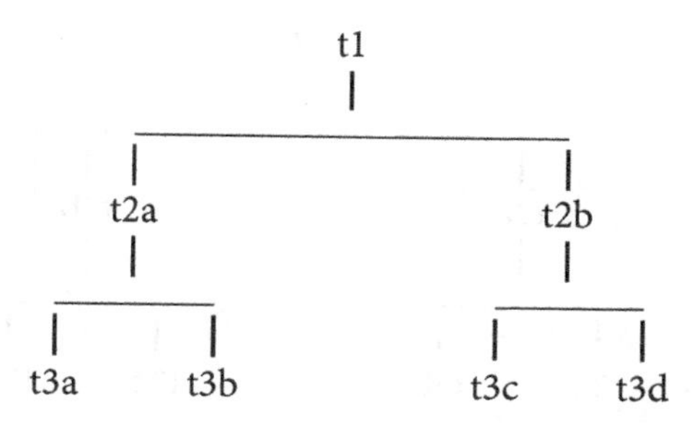

And suppose that space-battles are located at t3a and t3b but not at t3c and t3d. Then an utterance of "There will be a space-battle" located at t1 is open, since space-battles occur at nodes on some branches (some histories) that have t1 as a part but not on all. But an utterance of that sentence located at t2a or t2b is closed: it is closed at t2a because it is settled at t2a that there will be a space-battle, since every branch that has t2a as a part includes a node with a space-battle; and it is closed at t2b because it is settled at t2b that there will not be a space-battle, since no branch that has t2b as a part includes a node with a space-battle. But while it is not open at t2a or t2b whether there will be a space-battle, we can see that it is true at those nodes that it *was* open whether there will be a space-battle, since we can look to t1, which is in the past with respect to those nodes, and see from there futures where space-battles are occurring and futures where they are lacking. The alternative branches that will now never be our history still exist for us to look back on.

B-Theoretic branching salvages *talk* of the open future, but I deny that there is any genuine openness on this metaphysic. This seems like a metaphysic on which it is perfectly settled how things will be, you just do not know whereabouts you will be within reality. All those future possibilities are just there, and that never changes. If this is the correct metaphysics, then it is simply always settled that there are space-battles in reality; indeed, at t1 it is perfectly settled that there are space-battles temporally beyond us. It only comes out as unsettled whether or not there *will* be a space-battle because it is unsettled which of the branches that are temporally beyond us *will* happen. But that is not because reality is yet to determine which branch is the one that time is going to flow down; it is only because in talking about "our history" and "our future" we are choosing to only talk about a small corner of reality—the branch we happen to find ourselves upon. But there is nothing special about that branch, on this metaphysic—it is just that it is *our* branch. Genuine openness requires that *reality* be unsettled; openness should not arise just as a result of our parochialism.

So suppose we simply add a moving spotlight to the view. At t1, reality looks like this:

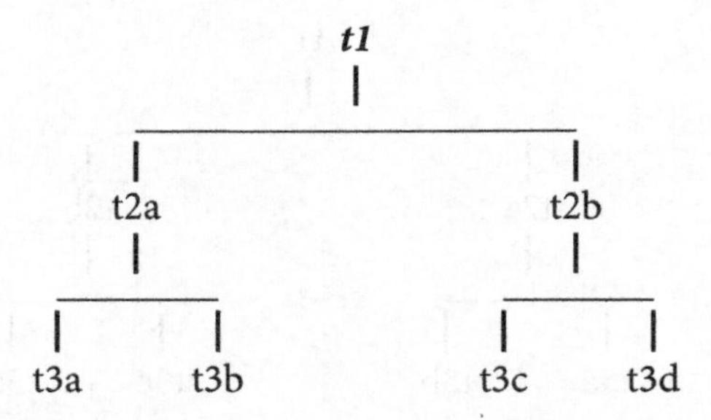

The bold and italic print in the diagram signifies not merely the node we happen to be observing reality from but rather what node is objectively present. "There will be a space-battle" is open simpliciter (not merely relative to a node), if reality is as above. And as time progresses, the spotlight moves up the tree, determining one branch amongst many as the unique history of the world as it progresses. So suppose time moves to t2a over t2b; reality now looks thus:

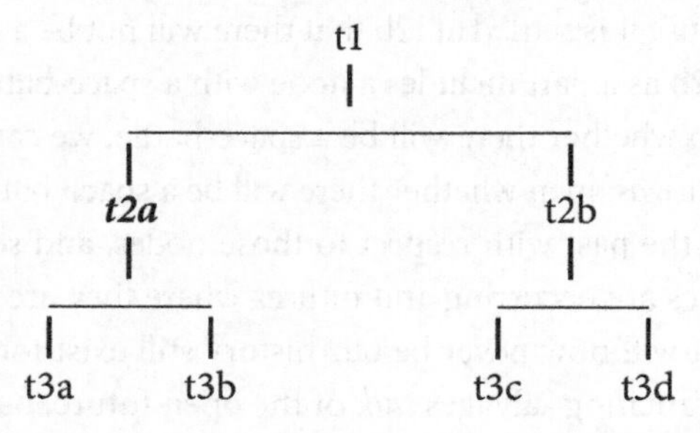

And now it is settled, simpliciter, that there will be a space-battle; although still true that it *was* unsettled whether or not there will be a space-battle.

I grant that there is genuine openness on this view. It is genuinely open which of the future branches the spotlight will move to. And there is a *metaphysical* reason for restricting your attention to one branch over all the others (namely, the one that the spotlight lights up) when making judgments about what *will* happen, and so since it is open *which* branch that is, it is open what *will* happen, as a result of how reality is as opposed to merely our parochial perspective on it. Nonetheless, I think this view ought to be rejected.

One problem is that it is hard to see why we should care about history unfolding one way rather than another, given this metaphysic. If all the ways that history could unfold are out there, and will always be out there, then of course I have *selfish* reasons to prefer to end up on a good one, but why would I have a *moral* reason to *bring about* a good history? It is open that there will be a space-battle, and you might not want there to be because it will lead to massive casualties. But while you have selfish reason to want to be on a space-battle-free branch, that is just like wanting to get out of the country when there is a war on. The war is still happening elsewhere, and the space-battle is still happening in reality: it is just not in your country/not on your branch. Good for you, but that doesn't matter morally!

For that reason, it seems to me that the best version of the branching view has the tree change as time progresses: not just in terms of which node is present, but in terms of what branches of the tree exist. The most natural option is that as time progresses, the branches which are ruled out as being a part of history cease to exist.[2] So, on this view, at time t1 things look as in the first diagram above. But assuming time then progresses to t2a rather than t2b, at time t2 things will look like this:

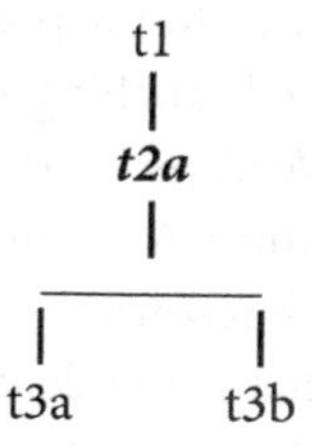

t2a's becoming present settles that t2b, t3c, and t3d never were and never will be present: they are ruled out as being parts of history, and as a result the branches containing them cease to be.

It is clear why, if this is the correct metaphysics of reality, I would have non-selfish reasons to want time to progress down one branch rather than another, for

[2] As in McCall (1994).

that makes a difference to how reality is as a whole. If the space-battle leads to nothing but destruction, then I should want time to progress down a space-battle-free branch, since reality as a whole will then not contain such destruction. That is a major advantage over the previous views. On the other hand, this view threatens to make our decisions have *too much* moral import, for now if we do something that results in time progressing down one branch, we condemn all those people on the rival branches to non-existence.

This version of the branching metaphysic has another problem. On this view, the present is the first node from which there is branching, since branches that *used* to diverge from earlier times ceased to be when those times became past. That means that this view on the face of it entails that time has been closed up until now, but from now on it is open. Look into the past and you will see only one history leading up to the present: it is only from now on that there are multiple ways for history to proceed. That is bizarre: of course it should be an open possibility that this is the first moment at which the future has been open, but it should not be guaranteed by the metaphysics that this is the case. To avoid such a conclusion, the believer in this view has to hold that what was the case cannot simply be read off from how the tree is before the present node; she needs to say that there is a brute tensed fact about how the tree used to be, in particular that in the past there were multiple ways history could have gone but did not.

This certainly makes the theory less elegant. But more importantly, it threatens to undermine the advantage promised by a branching metaphysic. If reality was open prior to now, despite reality not branching before the present node, then what is to rule out reality being closed in the future, despite reality branching after the present node? If there are tensed facts about how the branching reality *was* that are not given by how it *is* prior to the present, then what is to rule out there being tensed facts about how the branching reality *will be* that are not given by how it is beyond the present? And perhaps one of these tensed facts concerns what branch *will be* the actual future. As soon as the believer in branching admits brute tensed facts that give us the resources to say how the branching reality as a whole was or will be in a way that is not determined by how it *is* before or after the present node, she has to allow that it is an open epistemic possibility that there are (now) facts about which branch *will be* our history; and that is to allow that it is an open possibility that the future is not now open after all, despite the existence of these multiple branches after the present node. Now, of course, one can simply deny that there *are* such facts. But then the reason for the future's being open is not simply that there are these multiple branches; it is that together with there being no brute tensed facts as to which one will happen. But now openness is no longer being reduced to variation across the extant branches. The future's being open now involves these fundamental tensed facts. And so the branching view loses what looked to be its main advantage: to be giving an ontological account of what openness *consisted* in.

Here is another objection to any branching view: they cannot make sense of some possibilities that we ought to allow for. It seems possible that the future is open not simply in terms of *what* will happen, but whether *anything* will happen. That is, as well as it being open what will happen tomorrow, it may also be open whether or not reality will continue beyond tonight. So we should be able to make sense of there being multiple ways the future might unfold but also that it might not unfold any way because it is open—in exactly the same sense that each of those multiple futures is open—that this is the last time.

But how is the branching theorist to make sense of this? The problem is that the absence of further branches from a node does not represent the further open possibility that nothing will happen beyond that node, it simply represents the absence of further open possibilities. (Of course, one could stipulate the reverse as part of one's theory: but then one would make the metaphysics *guarantee* that every time might be the last time. But what we want is to be able to allow both for it being open that some time is the last time and for it being settled that it is not.) So look back to the first diagram above. Four branches diverge from t1, thus signifying that there are four ways the future might unfold from t1. What is signified by the absence of a fifth branch? That there are *only* four ways the future might unfold from t1, *not* that it might not unfold *any* way from t1 because t1 might be the last moment of history.

Now, of course you could have an extra branch on which nothing happens—a branch on which there are no space-battles, no people, no suns going supernova—but that is not the possibility of time ceasing, that is the possibility of time continuing but nothing happening. That might well be an extra way in which time might progress, but it is not the possibility we are trying to capture. Also, you can of course add to your branching diagram something which is *stipulated* to represent the possibility in question. So suppose we have:

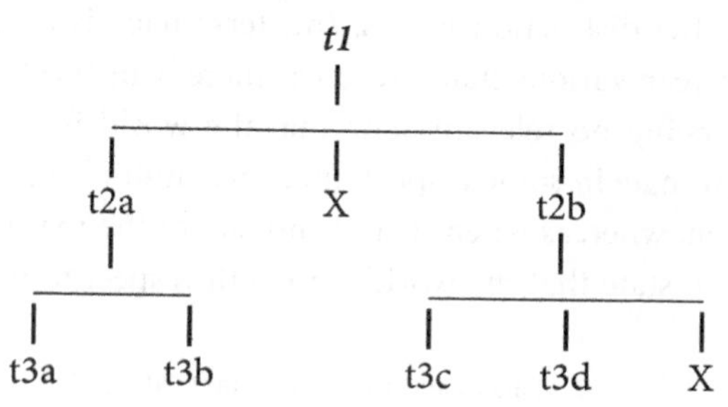

And we stipulate that a node with an X represents that the node leading to it could have been the last time. Then if things are as represented in the diagram, at t1 it was open that time would not continue past that point, and likewise at t2b, but at t2a that is not open: it is only open from t2a *how* time will progress, not *whether* it will.

Well, this is fine if the point of these diagrams is just to represent how things are without aiming to give a perspicuous picture of the underlying *metaphysics*; and that is fine if, for example, all we want them for is to do the semantics of future contingents. But our concern is with the metaphysics, and as such the above is entirely unsatisfying. What *are* the branches ending in Xs? They cannot actually be additional branches like the others, because that would be for time to continue in some form, not for it to have ended. So what in reality is actually represented by this addition to the representation? I cannot see a satisfying answer available to the branching theorist.

5.3 Indeterminacy versus No Fact of the Matter

So much for the branching ontology. Instead of trying to provide an ontological reduction of the unsettledness that applies to claims about what will happen, I think that we should take this unsettledness to simply be a brute feature of reality: that it sometimes be simply unsettled how reality is at its most fundamental level. That the open future consists in such brute unsettledness is a view that I have defended previously in joint work with Elizabeth Barnes.[3] But before I present the view, some clarification is needed as to what "unsettledness" amounts to. I actually think there are two phenomena that may each be thought of as a type of ontological unsettledness: there is the phenomenon of it being *indeterminate* whether something is the case, and there is the phenomenon of there being *no fact of the matter* whether something is the case. I think these are two distinct phenomena, and each can be used to characterize a different sense in which the future is open. In this section, I will make the case for treating these as two separate phenomena.

In a nutshell, the distinction is this. Indeterminacy is a matter of the world being poised between various states, whereas there being no fact of the matter is a matter of there being no relevant state that the world is in. That is, when the world is indeterminate in some respect, there are multiple states that the world is unsettled between, whereas when there is no fact of the matter for some respect, there is simply no state that the world is in with respect to that matter.

[3] Barnes and Cameron (2009, 2011).

There being no fact of the matter, then, is a kind of underdetermination. When there is no fact of the matter whether p, nothing worldly answers to the question of whether p. The worldly conditions for it being the case that p are absent as are the worldly conditions for it being the case that not-p. When it is indeterminate whether p, by contrast, it is not that both worldly conditions are absent, it is that it is unsettled which obtains. Indeterminacy is not the *absence* of worldly conditions, but a lack of specificity concerning *what* worldly conditions obtain.

There being no fact of the matter concerning something's being the case is, plausibly, incompatible with bivalence. When there is no fact of the matter as to whether p, nothing worldly answers to whether or not p, so p should be neither true nor false. It should not be true, because the worldly conditions required for the truth of p are absent, but nor should it be false, because the worldly conditions required for its falsity are also absent. So there being no fact of the matter excludes both p's truth and its falsity. Hence if there are some claims concerning which there is no fact of the matter, bivalence must be false.[4]

In drawing such a conclusion, I am relying on the weak principle that for p to be true, the worldly conditions for p's truth would have to obtain, and that for it to be false, the worldly conditions for its falsity would have to obtain. It follows from this very modest claim about how truth depends on reality and there being no fact of the matter regarding p that p is neither true nor false. This principle is much weaker than the truthmaker principle that demands for any truth that there be some things whose existence necessitates the truth in question; it is also weaker than the principle that demands that what is true supervenes on what exists and what fundamental properties they have. It says simply that truth and falsity depend on the world in the following sense: that propositions have truth- and falsity-conditions, and that for a proposition to be true/false the world must be as those truth-/falsity-conditions require it to be. That is fairly innocuous, and I will assume it throughout, and so I will speak of there being no fact of the matter whether p as giving rise to a failure of bivalence. Nonetheless, even such an innocuous principle will find its deniers. Patrick Greenough is one: he thinks that in cases where reality fails to make a proposition true or false, it will be a brute fact that it is one or the other.[5] The proposition will have a truth-value, but one that is entirely undetermined by how reality is. As I have said, I will assume this is

[4] I will assume throughout that 'Not p' is true just in case p is false, and so will take there being no fact of the matter whether p as ruling out the truth of p's negation. This is purely stipulative. One can introduce a "weak" negation such that "Not p" is true just in case p fails to be true. In that case, when there is no fact of the matter whether p, "Not p" will be true. Nothing hangs on this: so long as we are clear which negation we are using, there can be no cause for confusion.

[5] Greenough (2008).

false. But much of what I will say is compatible with it; instead of characterizing the difference between there being no fact of the matter and it being indeterminate as one between there being no truth-value and there being an unsettled truth-value, someone who holds Greenough's view should characterize it as the difference between there being a truth-value that is not determined by reality and there being a truth-value that is determined by reality but where it is unsettled *which* truth-value reality determines.

So given weak assumptions about how truth depends on reality, there being no fact of the matter concerning whether some proposition is true yields a failure of bivalence. Not so with indeterminacy. Indeterminacy does not arise from an absence of the relevant worldly conditions, but from unsettledness as to which obtain; hence, indeterminacy as to p does not exclude p's truth, it merely renders it unsettled as to *whether* p is true.[6] Many have of course been tempted to say that when p is indeterminate it is neither true nor false. But as Crispin Wright points out,[7] this does not seem to capture the phenomenon: adding a third truth-value does not help us capture the thought that some propositions are unsettled between the initial two truth-values. In what follows, then, I will assume that Elizabeth Barnes, and Barnes and J. R. G. Williams, are correct in saying that when p is indeterminate, it is determinately the case that either p is true or p is false, but it is indeterminate which, where this indeterminacy is a primitive unsettledness between ways the world could be.[8] And I will assume that this indeterminacy in truth-value results from an indeterminacy in ontology: that when p is indeterminate, it is indeterminate whether the worldly conditions required for the truth of p obtain or whether the worldly conditions required for the falsity of p obtain. It will be determinate that one or other set of conditions obtain (for if neither obtained, this would be a case of there being no fact of the matter whether p was true, not a case of its being indeterminate whether it was true), but indeterminate *which* set of conditions obtain.

Now, none of what has just been said is going to convince anyone who is committed to the view that indeterminacy *just is* there being no fact of the matter that there is another phenomenon in the vicinity here. After all, when distinguishing indeterminacy from there being no fact of the matter, I used the term "unsettledness": while there is no fact of the matter as to whether p excludes both the truth of p and the falsity of p, indeterminacy is not a matter of their

[6] I have been very influenced here by the picture of indeterminacy presented in Barnes (2006, 2010) and Barnes and Williams (2011).

[7] Wright (2003). See also Barnes (2010).

[8] Barnes (2006, 2010), Barnes and Williams (2011).

exclusion but of unsettledness as to which obtains. But the notion of unsettledness is hardly on a firmer footing than that of indeterminacy itself, and our current objector will likely simply interpret "It is unsettled whether p" as "There is no fact of the matter whether p," thus bringing us back to the "one phenomenon" view. I doubt there is any way of arguing definitively against the "one phenomenon" view from a neutral standpoint; but I do think the proposed difference between the phenomena is an intuitive one, and I think looking at some examples of where the phenomena potentially come apart can help to bolster those intuitions. For the remainder of this section I will try to pump your intuitions that there are two distinct ways in which reality may fail to settle some matter of fact, and draw out some of the consequences of each. In the following section we will use this to characterize two distinct senses in which the future may be open.

Let us start by looking at a very old purported case of indeterminacy. One question that occupied the scholastics was what Prime Matter was like. Prime Matter was postulated as that part of things which remains beyond their destruction. The scholastics were concerned with what is going on when, as it appears, something is destroyed and something else comes into existence: so, for example, a cat is killed, and is hence destroyed, and a cat corpse comes into being. Now, of course, one could deny this description of things and insist that there is one thing throughout that was a cat and is now a cat corpse; but to take this route is to head towards the Democritean view that nothing goes out of or comes into being, and this offended the common sense ambitions of the scholastics. So assume we have a genuine change in being occurring here. How can that be? It cannot simply be that one thing stopped and another thing came to be, thought the scholastics, because they were committed to the view that nothing could be created out of nothing: if something has come into being, it must have been made *from* something that already existed. But the cat corpse cannot have been made from the *cat* (so they thought), because the cat and cat corpse have never been around at the same time: by the time the cat corpse is here, the cat has departed the realm of being, so the cat corpse cannot have been made from the cat. So there must be something that was around before, when the cat was here, that is still around now and from which the cat corpse was made. This is Prime Matter: the underlying substance that formed both the cat and the cat corpse.

So what is Prime Matter like? Well, the problem is that there is nothing much you can say about what it is like. Prime Matter is meant to be the unchanging substance that underlies change. (The scholastics' thought seems to be that if Prime Matter itself could change, it could be corrupted and go out of existence, in which case there would need to be something underlying *it*, and so it would not be *Prime* Matter after all.) It must be the same no matter the quality of the thing it is underlying, so we can't say that it is like the thing it happens to underlie. This led to a lot of grasping at metaphors on behalf of the scholastics: that Prime Matter was "pure potentiality... free of all form, but open to all forms,"[9] that it was, as Averroes put it, "halfway between complete non-existence and actual existence."[10] It is not clear what that means. Perhaps the most intelligible account is offered by Peter Auriol, who claims that Prime Matter is simply of indeterminate character:

> Prime Matter has no essence, nor a nature that is determinate... it is pure potential, and determinable, so that it is indeterminately and indistinctly a material thing.... It is not determinately any of the beings in the world—such as stone, earth, and so forth—but it can be determined so as to be stone, earth, and so forth.[11]

Why invoke indeterminacy, rather than just saying that it outright lacks the qualities in question? That is, why not simply say that Prime Matter is *not* stone, *not* earth, *not* round, *not* red, etc., rather than claiming that it is indeterminate in quality? The scholastics' thought seems to be that if it is simply *false* that it has any of these qualities, that is too close to saying that it is nothing at all. It can't be that it is *no* way—that it is simply *lacking* in quality—for everything that is is *some* way or other: rather, it must be indeterminate in quality.

This is a case in which I think we would be better using "no fact of the matter" rather than "indeterminacy." We are dealing here with a case of ontological underdetermination: nothing in the world is answering to questions about the quality of prime matter. If we look to see whether the conditions for Prime Matter being red obtain, we find in the world nothing that corresponds to it being red and nothing that corresponds to it not being red: there are simply no color facts concerning Prime Matter. Contrast this with another scholastically influenced case of indeterminacy concerning changes in being.

Elizabeth Barnes[12] has argued that there must be cases of indeterminate existence if (1) time is gunky, and (2) changes in existence are underwritten by Aristotelian events of "coming to be." The thought is this: suppose A exists, but did not always exist. By (2) there must have been an event of A's coming to be. Given (1), this event was not instantaneous. *No* event is instantaneous, because there are no instants: there are only extended temporal regions, and hence every event occupies some extended temporal region. In that case, there can be no

[9] See the remarks of Eustachius and Burgersdijk, quoted in Pasnau (2011, p35).

[10] Quoted in Pasnau (2011, p38).

[11] Quoted in Pasnau (2011, p39).

[12] Personal correspondence.

instantaneous transition from A's not existing to its existing: instead, there had to be an extended process during which A came to be. Now what should we say about the status of A during this event? Not that it exists, says Barnes, for then it would have already come to be. But not that it does not exist, either, she argues, for there is a *difference* concerning A's being from before it started coming to be and now when it is coming to be. So we should say, says Barnes, that while A is coming to be, its ontological status is indeterminate: "A exists" is neither determinately true (as it will be once it has come to be), but nor is it determinately false (as it was before it started to come to be)—it is indeterminate whether it is true or false.

One might worry that Barnes' reasoning about this case cannot be right because, if it were so, it would give rise to a troubling higher-order indeterminacy.[13] If there can be no instantaneous transition between A's not existing and its existing, surely the same goes for the transition between A's not coming to be and its coming to be. And if the non-instantaneous transition between A's not existing and its existing is a period wherein it is indeterminate whether A exists, then the non-instantaneous transition between A's not coming to be and its coming to be should be a period wherein it is indeterminate whether A is coming to be: that is, a period wherein it is indeterminate whether it is indeterminate whether A exists. And so on.

Two points in reply. First, it is not clear that this resulting higher-order indeterminacy would be so terrible. In a gunky time, there is no problem with fitting all these periods in: it is perfectly consistent that there be a duration of time during which it is indeterminate whether A exists, which is bordered by a smaller duration of time during which it is indeterminate whether it is indeterminate whether A exists, which is itself bordered by a duration of time which is smaller still during which . . . and so on, ad infinitum. A gunky time can easily accommodate the infinitely many levels of increased unsettledness, so the case would need to be made that this is indeed problematic. Secondly, the regress can be resisted if desired. Note that Barnes' case relies on the Aristotelian thought that changes in being are underwritten by events of coming to be. That is crucial: that existence facts change and time is gunky is not enough to support a claim of indeterminacy. The reason is that even in a gunky time, we can still say that A fails to exist in such-and-such a temporal region, T, and exists in every temporal region that is later than T. In effect, we can model instantaneous change even when there are no instants. Barnes' case gets going because she demands that there be an event that underwrites the change: this event must occur during an extended interval of time (since all intervals of time are extended), and it is this

[13] This objection was suggested by Matti Eklund.

that forces us to ask: what is A's status during that interval? For the regress to get going, then, we must demand—as well as an event of A's coming to be—an event of the event of A's coming to be coming to be. But that could be resisted in a non-ad hoc manner: one could hold that when a substance comes into the world there is an event of its coming into being, but deny that the same holds of events. Substances *exist*, events *happen*; if you think that marks a genuine ontological distinction, it is not crazy to think that events do not come into being in the same way substances do. If one took this line, then for every temporarily existing substance there will a period of time during which it is indeterminate whether it exists, but one is not forced into saying that there is a period of time during which the event of its coming to be indeterminately exists; hence one is not forced into saying that there is a period of time during which it is indeterminate whether it is indeterminate whether the substance exists, and so there is no pressure to accept higher-order indeterminacy.

Now, intuitively, I think, the case of A's existence while it comes to be is very different from the case of the quality of Prime Matter. In the former case, there *is* something worldly we can look to that corresponds to the question of whether A exists: it is the event of A's coming to be. It is just that we don't know—and nothing settles—whether that portion of reality counts *for* A existing or *against* it. The ontology is there, it's just unsettled whether it corresponds with the proposition or its negation. Whereas in the case of what Prime Matter is like, it's just that reality is utterly lacking in anything that counts towards or against any claim about what it is like.

In the Prime Matter case, every judgment we could make about its quality seems absolutely wrong. Suppose we ask about the shape of Prime Matter. Prime Matter is matter *independent* of qualities, so it seems entirely mistaken to say that it is square, or round, etc. But nor do we want to say simply that it is not round, not square, etc.—for if we say that for every shape that matter could be, as we would have to, then we are saying that Prime Matter has no shape, and that is wrong as well, for being shapeless is just another way to be, and Prime Matter is matter *independent* of the ways for matter to be. So *everything* we can say about its shape is wrong, and so we have to make no verdict either way: we conclude that there is simply no fact of the matter concerning the shape of Prime Matter. Nothing in the world speaks to such a question.[14]

[14] Contrast the case of Prime Matter with the case of abstract objects like, e.g., the number 2. It seems absolutely right to say that 2 is not round, not square, not triangular . . . etc., and to conclude that 2 has no shape: that it is shapeless. In this case, it is only the positive verdicts that are ruled out, with the negative verdict that 2 is simply lacking in shape being the right result. The peculiar situation of Prime Matter arises because the negative verdict is equally bad as the positive ones. This

With Barnes' case of coming to be, it is not that everything we can say about the existence of A seems wrong: it is that there are multiple candidate options such that they each seem equally good. Nothing rules out A's existing or A's not existing in the way that the very nature of Prime Matter rules out its being square but also its being some shape other than square. It is not that "A exists" and "A doesn't exist" are both bad things to say—the problem is that they are both kind of good things to say but, because they are *equally* good things to say (and we can't say them both), they are also kind of bad things to say.

Each case puts us in a quandary about what to say, but for very different reasons. In one case, everything we could say seems wrong; in the other, none of them seem outright *wrong*, but they seem equally good, and we cannot say them all. I think we should take the appearances at face value: in the first case, everything we could say seems wrong because reality fails to speak to the question one way or the other—thus reality vindicates neither any positive verdict nor any negative verdict; in the second, each of the options seems equally good, but none seem wrong, because reality *is* speaking to the issue, but it is not settled *how* it speaks to it—thus reality vindicates some verdict, but it is unsettled *which*. The former is a lack of a fact of the matter, the latter indeterminacy. And so I think that Barnes' case is a case of indeterminacy proper, whereas the Prime Matter case is a case of there being no fact of the matter. And so, when we are to make a judgment about the truth-value of "A exists" while it is coming to be, we should say that it is *either* true or false, but that it is indeterminate which, whereas we

is why we should say that reality determines that 2 lacks shape but that it simply fails to determine any fact regarding the shape of Prime Matter; hence, while it is simply false that 2 is round, there is no fact of the matter whether Prime Matter is round.

Contrast also the case of a vague object which is indeterminate in shape by virtue of it being indeterminate what its parts are. If each of the Xs are part of A then A's shape is one way—determined by the locations and shapes of each of the Xs—but if only a sub-plurality of the Xs, the Ys, are part of A then A's shape is some other way—determined by the locations and shapes of the Ys. So if it is indeterminate whether all the Xs or only the Ys are parts of A, then it is indeterminate what A's shape is. But the case of A's shape is very different from the case of Prime Matter's shape. The world is speaking to A's shape: the mereological and locational facts concerning A, the Xs and the Ys speak to. But it is unsettled how the world speaks to A's shape, because the mereological facts are unsettled: reality does not settle whether the Xs that are not Ys are parts of A, and this matters to A's shape. But because the world speaks to the issue, plenty of claims about A's shape are ruled out: it is determinately not the case that A has the shape of some thing that has the Xs as parts but that has lots of other things as parts in addition, and it is determinately not the case that A has the shape of a single one of the Xs. Reality is speaking to the issue, just not with complete specificity—hence, this is a case of indeterminacy. By contrast, reality not only does not give us a unique best verdict with respect to the shape of Prime Matter, it fails to rule out any verdict whatsoever—hence, this is a case of there being no fact of the matter.

should say of "Prime Matter is red" that there is no fact of the matter concerning this claim: it is *neither* true nor false.[15]

If the foregoing is right, it has consequences for the question of how ontology lines up with claims of determinacy and indeterminacy. Consider two apparently rival views. Patrick Greenough argues that determinate truth is truth that is made true.[16] So if there is a truthmaker for p/not-p, then it is determinately true/false that p. Indeterminacy arises in the absence of a truthmaker: so if both p and not-p lack a truthmaker, then it is indeterminate whether p. A similar view is held by Jessica Wilson, who argues that indeterminacy arises when a determinable is instantiated without any further determinate of that determinable being instantiated.[17] So, for example, if Ball instantiates *is colored* without instantiating any of the particular colors that determine that determinable, then it will be indeterminate what color Ball is. This can be seen as a special case of Greenough's position, taking a particular stance on what the relevant truthmakers are: namely, that they are determinates and determinables.

By contrast, I have defended the view that all truths are made true, but that it can simply be indeterminate *which* potential truthmaker exists.[18] So if it is indeterminate what color Ball is, it is not—as Greenough and Wilson would have it—that there is no truthmaker for "Ball is red," for example, but rather it is indeterminate *what* truthmaker exists. That is, there are a bunch of potential things that could exist, one would make it true that Ball is red, one that Ball is orange, etc., and it is determinate that exactly *one* of these things exists, but it is indeterminate *which* one of them exists. A similar view was defended by Barnes who argues that cases of semantic indeterminacy are those in which it is perfectly

[15] A case that is a lot like that of Prime Matter is that of Meinong's incomplete objects. Meinong (1915, §25) thinks there is such a thing as *the triangle*: this thing has three sides, but it is neither equilateral, isosceles, nor scalene. The nature of this object is simply not specific enough to determine any truth concerning the relative lengths of the three sides that this thing has. This is a clear case of reality failing to speak to an issue: there is nothing in the world to speak to the truth or falsity of any claim concerning the relative lengths of the lines of the triangle. The nature of *the triangle* underdetermines what it is like in this respect. The positive claim "Each of the triangle's sides are the same length" is ruled out by the nature of the triangle, since *the triangle* is not an equilateral triangle, but so is the negation of that claim ruled out by its nature, since *the triangle* is also not an isosceles or scalene triangle: it is just a triangle. Contrast this with a case where there is a particular triangle where it is indeterminate whether the third side is the same length as the other two (which are determinately the same length): in this case the nature of this particular triangle *is* speaking to whether it is equilateral or isosceles; it is just that it is indeterminate which it is. Reality *is* speaking to the truth or falsity of "This triangle is equilateral"; it is just that it is indeterminate which truth-value it has, because it is indeterminate how the relevant portion of reality is. There is no fact of the matter whether Meinong's incomplete object *the triangle* is equilateral, whereas it is indeterminate whether this particular complete triangle is equilateral.

[16] Greenough (2008). [17] Wilson (2013). [18] Cameron (2011).

settled what exists but it is unsettled what those things make true, whereas ontic indeterminacy arises when it is perfectly settled what would be made true by some things, but it is unsettled which things exist.[19]

So we have two apparently very different accounts of how ontology interacts with indeterminacy: one view sees indeterminacy as the result of a gap in ontology—it is perfectly settled what there is, and indeterminacy is a result of the settled absence of relevant truth- or falsity-makers; the other view sees indeterminacy as the result of an unsettledness in ontology—it not being settled what there is, hence it not being settled whether what there is makes it true or false that p. But if what I have said above is correct, I do not think we should see these views as rivals, but rather as characterizing two distinct phenomena: the Greenough/Wilson view should be viewed as a proposal about what reality is like when there is no fact of the matter, the Barnes/Cameron view as a proposal about what reality is like when it is indeterminate. An *underdetermination* in ontology—the absence of a truthmaker for both a claim and its negation—leads to there being no fact of the matter concerning what that claim says: it is neither true nor false. A lack of specificity in ontology—it being unsettled just what exactly there is, hence unsettled whether a truthmaker for some claim exists—leads to it being indeterminate whether that claim is true. Two distinct phenomena; two distinct ways in which ontology may fail to settledly determine whether some claim is true or false.

So how do we know when we are dealing with indeterminacy and when we are dealing with there being no fact of the matter? Well, in many cases I think it is intuitively clear, once we have the distinction in mind. Indeterminacy arises when we have a transition between two states, and the sense of there being no point such that *it* is the point of transition. A sorites series is a paradigm case, but not the only case. There need not be anything soritical about Barnes' case of coming into being, but there is still the sense of a transition between states (from non-being to being), and the sense of there being a time at which it is unclear which state we are in (namely, during the process of coming to be), and that this unclarity is not the kind that could be resolved if only we knew more. But this case doesn't hang on the plausibility of claims like "If A doesn't exist at time t, it doesn't exist at time t+Δ, for short Δ." In paradigm cases of there being no fact of the matter, by contrast, there is not the sense of transition between rival states, but rather the sense of there being no such states at all. So consider a sometimes purported case of indeterminacy: the continuum hypothesis. If you are attracted to a view on mathematical truth such that it cannot outstrip what is in principle

[19] Barnes (2010).

provable, then you might be hesitant to say that reality has settled the truth or falsity of the continuum hypothesis and that it is merely epistemically inaccessible to us, thinking instead that reality simply does not settle this issue. But there is no sense here of reality being poised between the two states: it is not that mathematical reality is rich enough to speak to the continuum hypothesis but that it is unsettled how this rich mathematical reality is; rather, it is that mathematical reality just is not extensive enough to speak one way or the other towards the continuum hypothesis. So it is not indeterminate, there is simply no fact of the matter.

Other cases, however, resist immediate classification on intuitive grounds, and whether or not they should be classed as cases of indeterminacy or as the lack of a fact of the matter will depend on broader theoretical considerations. Here is one helpful consideration. There can only be no fact of the matter concerning a claim in some domain if it is not sufficient for the truth of the negation of claims in that domain that the conditions required for the truth of those claims fail to be met. For there being no fact of the matter regarding p, as I have characterized it, requires that both the worldly conditions for the truth of p and the worldly conditions for its falsity be absent. In that case, if there can be no fact of the matter as to whether p, it cannot be the case that the conditions for the falsity of p simply *are* that the conditions for its truth be absent. There being no fact of the matter demands two things, one of which is the absence of the conditions for p's being true: if the absence of such conditions is sufficient for p's falsity then part of what is required for there to be no fact of the matter ensures that there *is* a fact of the matter (namely, that p is false), and hence there being no fact of the matter makes an inconsistent requirement on the world, and so it follows that it cannot arise that there is no fact of the matter regarding p.

So it cannot arise that there is no fact of the matter regarding p when the falsity-conditions of p are simply that the truth-conditions for p fail to be met. It can only arise that there is no fact of the matter as to p when p has truth-conditions and falsity-conditions that can each fail to be met. This is what is going on in the continuum hypothesis example. The truth of the continuum hypothesis requires that the realm of numbers extends so far and is a certain way, its falsity that it extends so far and is some other way; either way, for it to have a truth-value requires of the world that the realm of numbers extends that far, and so if this simply is not the case—if a certain strict finitism is true, for example—then neither condition is met, and so there is no fact of the matter regarding the continuum hypothesis.

Often, whether you will see it as even conceptually possible that there be no fact of the matter concerning some claim will depend on your broader theoretical commitments. Consider, for instance, an example where the possibility of there being no fact of the matter depends on certain meta-ethical claims. Take "It is morally impermissible to Φ." The orthodox view, I take it, is that one is permitted to Φ iff there is no moral obligation to refrain from Φ-ing: things are permitted in the absence of obligations to the contrary. If this is so, then the truth-conditions of "It is morally impermissible to Φ" are that there is an obligation to refrain from Φ-ing, and the falsity-conditions are simply that there is no such obligation. In that case, it cannot arise that there is no fact of the matter as to whether it is impermissible to Φ, as this is a case when the conditions required for falsity are just that the conditions required for truth fail to obtain, so it cannot be that neither condition is met. But suppose someone held the non-orthodox view that having a permission to Φ makes some positive requirement on the world, not merely that there be no conflicting obligation. To fix ideas, let us suppose that permission to Φ requires God's sanction, while an obligation to refrain from Φ-ing requires God's commandment not to Φ. Now it is no longer the case that the falsity of "It is morally impermissible to Φ" requires simply that its truth-conditions fail to be met: its truth requires God's prohibition on Φ-ing, but its falsity requires not simply that God fails to prohibit it but that He sanctions it. But maybe He does neither; maybe He just stays silent on the question of Φ-ing—in which case, there would be no fact of the matter as to whether it is morally impermissible to Φ. The impermissibility conditions are lacking, but so are the permissibility conditions, and so reality simply does not speak to the issue either way. Now, that is not a popular meta-ethical view,[20] but it illustrates how the possibility of there being no fact of the matter can depend on serious philosophical issues elsewhere.

Here is another example of how your theoretical considerations elsewhere can affect whether you can admit the intelligibility of there being no fact of the matter concerning something. Consider an existential claim, "There is an F." Perhaps we are hesitant to attribute a definite truth-value to this claim for some reason. Should we judge it as indeterminate, then, or as there being no fact of the matter?

[20] The possibility of *legal* gaps—that it is neither legally permitted to F nor legally forbidden to F—is a live issue in the philosophy of law. (See, e.g., Alchourrón and Bulygin (1971, 20ff) and Soeteman (1997).) If there are such legal gaps, then there will be no fact of the matter whether one is legally permitted to F. It seems to be generally assumed, though, that even if legal gaps are possible, moral gaps in the sense outlined in the text are not. I see no reason to deny their intelligibility, however. (Thanks to Ralph Wedgwood and Krister Bykvist for guiding a wayward metaphysician to the relevant moral philosophy literature here!)

Well, can it even arise that there is no fact of the matter whether there is an F?[21] It depends, I think, on whether you are a truthmaker maximalist.

On the face of it, there *cannot* be no fact of the matter whether there is an F, because existential claims are amongst those where the conditions required for their falsity are just that the conditions required for their truth be absent. How does the world need to be for it to be true that there is an F? That there be an F. How does it need to be for that to be false? Just that the conditions for truth not obtain: that there be no F. So, seemingly, it cannot arise that both the conditions for its truth and the conditions for its falsity fail to be met, since the conditions for its falsity just *are* that the conditions for its truth fail to be met. So there cannot be no fact of the matter as to whether there is an F; although it could, for all that has been said, be indeterminate.[22]

However, while that is the most intuitive view, some—the truthmaker maximalists—will disagree: thus where you stand on the metaphysical debate concerning truthmakers will affect whether you see there being no fact of the matter concerning existential claims as even an option. For the truthmaker maximalist, the falsity-conditions of "There is an F" are not simply that there be no F: the falsity of the existential claim requires there to be a totality state of affairs (or similar entity) that makes it true that all the non-Fs are *all the things that there are*. In that case, they can make sense of there being no fact of the matter as to whether there are Fs. This would arise when reality fails to include Fs in its ontology but also fails to include anything like a totality state of affairs: the truth-conditions of "There are Fs" are thus not met, but neither are its falsity-conditions. Reality, in lacking a totality state of affairs, fails to make *any* existential claim false, it only makes some true. The world says there are such-and-such things, but it never says *that's it*, so it never settles that there *aren't* so-and-so things. Now, of course truthmaker maximalists tend to think that such a situation is metaphysically impossible: they (mostly[23]) hold that it is necessary that there is some totality state of affairs or other, and therefore hold that it is

[21] Let us stipulate that "F" is not a vague predicate, so that any indeterminacy/lack of a fact of the matter concerning whether there is an F is genuinely *existential* indeterminacy/lack of a fact of the matter.

[22] Theodore Sider (2003b) argues that indeterminate existence is impossible; but see Barnes (2013) and Woodward (2011) for responses.

[23] The sole exception here, as far as I know, is Stephen Mumford (2007a, 2007b), who agrees with the maximalist that every truth requires a truthmaker but agrees with the opponents of maximalism that there is nothing like a totality state of affairs, and as a result denies that negative existentials are ever true. So the conditions for the truth of "There are no Arctic penguins" are not met, because there is no totality state of affairs to ensure that the ontological inventory is completed (so as far as reality is concerned, there could still be an Arctic penguin somewhere), but of course the conditions for its falsity are not met either, since that would require reality to provide an Arctic penguin, which

necessary that if there are no Fs, then there is a totality state of affairs that makes it true that all the non-Fs are *all* the things that there are, and hence makes it false that there is an F. Thus they will probably hold that it is metaphysically impossible for there to be no fact of the matter whether there are Fs. But nonetheless, they should concede the *intelligibility* of the claim, even if they rule out its metaphysical possibility; for surely there is nothing *incoherent* about supposing that there is no totality state of affairs, even if there must be one. This is in contrast to the denier of truthmaker maximalism, who will not even be able to make sense of the claim that there is no fact of the matter in this case.

To sum up what has been argued for in this section: there are two potential kinds of phenomenon that one might call cases of the world being unsettled. It can be indeterminate whether things are thus-and-so, or there can be no fact of the matter whether things are thus-and-so. There can only be no fact of the matter as to whether p is the case if p's truth and its falsity both place a demand on the world that can fail to be met—so the conditions for p to be false cannot simply be that it fail to be true. Whether this is so in any particular case will often depend on your broader theoretical commitments.

5.4 Two Kinds of Openness

I said that we should characterize the thesis that the future is open as a claim about the world being unsettled in a certain respect. It will come as no surprise, given the above, that I think there are two ways to characterize the open future thesis: as the thesis that (at least some) future contingent statements are indeterminate in truth-value,[24] or as the thesis that there is no fact of the matter regarding the truth-value of (at least some) future contingents.

The thesis that there is no fact of the matter concerning future contingents sits well with the growing block metaphysic. As I characterized it above, for there to be no fact of the matter whether p requires an underdetermination in ontology: nothing in reality speaks either to p's truth or to its falsity—both p's truth-conditions and p's falsity-conditions simply fail to be met. The very possibility of that requires that p's falsity-conditions not be simply that its truth-conditions do not obtain; there must be an independent understanding of what is required

it does not do. So as I am using the terminology, Mumford should hold that there is no fact of the matter whether there are Arctic penguins.

[24] This is the version of the open future thesis that I defended with Elizabeth Barnes in Barnes and Cameron (2009, 2011). Those interested in further details of the view of the open future defended in this chapter should check out those papers.

for p to be false other than that it simply not be made true. This, I think, is how the growing blocker will naturally view things. Take a future contingent claim like "There will be a space-battle." What is required for it to be true is that the future be a certain way—that is, that it be one containing space-battles—and what is required for it to be false is that the future be a different way—that is, that it be one lacking in space-battles. But if, as the growing blocker thinks, there simply *is* no future, then neither condition is met. The case is exactly analogous to the continuum hypothesis case mentioned above. The truth of the continuum hypothesis demands that mathematical reality be extensive in a certain respect while its falsity demands that mathematical reality be extensive in a different respect: for there to even be a fact of the matter, then, demands that there be an extensive mathematical reality in the first place to be one way or the other. Similarly, the growing blocker is most naturally taken as thinking that the truth of claims about what will happen demands that temporal ontology be extensive in a certain way (that it includes a future in which what is predicted by the claim does indeed happen) and that its falsity demands that temporal ontology be extensive in a different way (that it includes a future in which what is predicted by the claim does not happen). For there to be even a fact of the matter, then, demands that our temporal ontology be appropriately extensive in the first place: that it includes a future. Since the growing blocker denies that our temporal ontology is so extensive—she denies that there is any future whatsoever—she ought to deny that there is any fact of the matter regarding the truth-value of this claim about the future. It is neither true nor false.

I say that this is the "most natural" way to interpret the growing blocker. Certainly, the growing blocker is not *forced* into saying that there is no fact of the matter concerning predictions of what will happen. The growing blocker is not even forced into saying that the future is open in any sense whatsoever: she could hold, for example, that while there is no future ontology, there are brute facts about what will happen. Or she could accept the open future as a matter of indeterminacy as opposed to there being no fact of the matter, by saying that it is indeterminate which of these brute facts about what will happen obtain. But such options, while consistent, are unattractive, for they treat claims about what did happen and claims about what will happen as being sensitive to different kinds of features of the world in a way that is ad hoc. The growing blocker treats claims about what did happen as being sensitive in some way or other to past ontology; in that case, due to considerations of parity, claims about what will happen should be sensitive to future ontology in just the same way. In which case, since there simply is no future ontology on her view, she should think that there is no fact of the matter concerning any (contingent) claim about what will happen. So while it is *consistent* for the growing blocker to accept a different, or even no, account of the open future, I think that by far the best thing for the growing blocker to say is that the future is open in that any future contingent lacks a truth-value.

By contrast, I think the moving spotlight view sits naturally with the thesis that if the future is open then it is a matter of future contingents being indeterminate: that for any prediction about how things will be, it is (determinately) either true that things will be that way or false that things will be that way, but it is (sometimes, at least) indeterminate which. The moving spotlighter agrees with the growing blocker that claims about what was the case are sensitive, in some respect, to past ontology and so, again due to considerations of parity, predictions about what will happen should be sensitive in the same way to future ontology. In that case, given that the moving spotlighter believes in future ontology, she believes in the relevant portion of reality that should speak to the truth or falsity of any claim about how things will be. Unlike the growing blocker, she believes that our temporal ontology is extensive enough to speak to the prediction, one way or another. *Prima facie*, then, either that ontology is as the claim predicts, in which case the prediction is true, or it is not, in which case it is false. One way or another, there is a fact of the matter. But this is perfectly consistent with it being indeterminate *which* fact of the matter obtains. The commitment of the moving spotlight view is just that there *is* future ontology; it is no commitment of the moving spotlight view that it is determinate what that future ontology is like. If there are future entities but it is indeterminate what they are like, this allows that it is indeterminate what future contingent claims those future entities make true. Thus there can be claims about what will happen that, due to the indeterminacy in how the future is, are neither determinately true nor determinately false (although it is determinate that they are either true or false). Reality is speaking to the issue—it contains the relevant ontology to make true the claim or its negation—but it is indeterminate whether reality is speaking to the truth or to the falsity of the claim, because it is indeterminate what the relevant ontology is like. This results in what Elizabeth Barnes and I called the "growing cloud of determinacy" model.[25] What will happen is indeterminate because how future ontology is is indeterminate . As time progresses, it is not that future things come into being, as the growing blocker says, but rather that how things are gets more *settled*: claims that were indeterminate, because they characterized future ontology, become determinately true or determinately false as that future

[25] Barnes and Cameron (2009, pp305–6).

ontology becomes present ontology and, as a result, the indeterminacy in how those things are gets resolved.

So while the growing block metaphysic naturally sits with an account of openness in terms of there being no fact of the matter concerning future contingents, the moving spotlight naturally sits with an account of openness in terms of future contingents being indeterminate. Presentism, by contrast, can happily sit with either. The presentist thinks that truths about what was the case are not sensitive to past ontology: rather, such claims are sensitive to either how present ontology is, or to what brute historical facts obtain. By parity, then, she should say that predictions about what will happen are sensitive either to how present ontology is, or to what brute future-directed facts obtain. But while it is a commitment of the moving spotlight view that there is future ontology, and a commitment of the growing block view that there is no future ontology, presentism is simply silent on whether there are the relevant brute future-directed facts, or whether present ontology has a rich enough nature to speak to the truth-value of a future contingent. The presentist can hold that there are brute future-directed facts, say, but that it is indeterminate which such facts obtain, thus accepting openness as indeterminacy; but she can equally well hold that there simply are no brute future-directed facts (or any relevant properties of presently existing things), and so reality simply does not speak for or against some future contingent, thus accepting openness as there being no fact of the matter. Neither option sits any better with presentism, per se, so the presentist has more choice here than her opponents. Not that this is much of an argument for presentism, though; as I will argue in the next section, there is independent reason to prefer an account of openness in terms of indeterminacy over an account of openness in terms of there being no fact of the matter. Given the above, this gives us an argument for the moving spotlight or presentism over the growing block.

5.5 Openness as Indeterminacy

I think there are reasons to prefer an account of openness as indeterminacy over an account of openness as a lack of fact of the matter. That is, if you think the future is open with respect to whether or not there will be a space-battle, you ought to think that this is a matter of "There will be a space-battle" being, determinately, either true or false, but it being indeterminate whether it is true or false, rather than thinking that it lacks a truth-value. If this is right, we should accept an ontology of time whereby reality is rich enough to speak to the truth or falsity of future contingents, but where it can be indeterminate how reality is in the relevant respect, rather than an ontology whereby reality is simply lacking the relevant ontology. That means we ought to reject the growing block and accept either presentism or the moving spotlight. (So combined with the arguments in section 4.4 for the need for non-present entities, this gives us an argument for the moving spotlight.) Here are three reasons for preferring the indeterminacy account of openness.

Reason 1: The Indeterminacy Account is More Flexible as to What it Can Allow to be Open and Fixed

The first reason is not actually directed against the no fact of the matter account of openness per se, but to that account combined with the growing block metaphysic.

It is no part of the concept of the open future that it be all or nothing: that is, that for *every* future contingent claim, p, it be open whether things will be as p says, or that for every future contingent claim, p, it be fixed whether or not things will be as p says. It is a conceptual possibility that the future is open in some respects and fixed in others; that is, that there is some future contingent, p, such that it is open whether things will be as p says, and some future contingent, q, such that it is fixed that things will (not) be as q says. As such, I think it is a *pro tanto* virtue of an account of openness that it allows such flexibility: that while it allows that the future is open in every contingent respect, it also allows that it be open in some but not all contingent respects. Ideally, our account of *what it is* for the future to be open should remain silent on the extent of *how* open the future is.

This, I think, speaks against the growing block view. It looks inevitable that the growing blocker has to hold that the future is open with respect to every contingent way it might be. After all, there is simply *no* future ontology according to that metaphysic. But, presumably, every future contingent is sensitive to how the future will be—that is what makes it a *future* contingent—and it could be true or false depending on how exactly the future will be—that is what makes it a future *contingent*. So if there simply is no future to make things one way or the other, there will be no fact of the matter as to whether things will be as that future contingent predicts. There seems to be no room for the growing blocker to say that *some* future contingents are fixed: what would make it true (or false) that things will be that way, given that the relevant features of reality that would speak to the issue—future ontology—are simply lacking?

By contrast, the moving spotlight view, with its account of openness as indeterminacy, is compatible with both a maximally extensive and a more restricted account of the *extent* of openness. Openness, on this view, arises because it is indeterminate what the future is like. This is completely silent on the extent of that indeterminacy. How indeterminate is the future ontology that the moving spotlighter believes in? Perhaps maximally so: that is, for every contingent claim characterizing future ontology it is neither determinately true nor determinately false that the future entities the moving spotlighter believes in are that way, in which case for every future contingent, p, it will be open whether or not things will be as p predicts. Or perhaps the extent of indeterminacy is simply that one of the future entities, A, is indeterminately F, with every other claim about how future entities are being determinately true or determinately false, in which case all that is open are the future contingents whose truth-value depends on whether or not A is F, with every other future contingent being fixed one way or the other. Or perhaps it is somewhere in between. The moving spotlighter is free to give whatever account she likes of the extent to which the future is open. I think that is a virtue of the theory.

Reason 2: The Indeterminacy Account Satisfies Intuitions about Retrospective Assessment

As John MacFarlane has emphasized,[26] the openness of the future seems compatible with the ability to make retrospective assessments about the truth or falsity of predictions about how things would turn out. So while I might think it now open whether or not my prediction that it will rain tomorrow is true, once tomorrow arrives and it is indeed raining I am not only inclined to think that it is true now that it is raining, but I am also inclined to think that this reveals that yesterday's prediction that it would rain *was* true when I made it. I take the current weather as sufficient grounds for concluding that the earlier prediction was in fact true.

It is hard for anyone who takes openness to consist in there being no fact of the matter to agree with this. When I consider predictions made in the past about how things would be at the time that is now present, I seem forced into saying that they were neither true nor false. After all, it was open how things would turn out, and openness consists in there being no fact of the matter as to what will happen. So I should say that while it is raining now, yesterday's prediction that it would be raining today was not true, because it was open what the weather would be like. That just seems wrong: if things are a way, then predictions made to the effect that things will be that way seem true.

MacFarlane himself attempts to reconcile the openness of the future with this data concerning retrospective assessment by adopting a relativist semantics

[26] MacFarlane (2003).

whereby tensed claims have a truth-value only relative to a context of assessment.[27] The idea is that the very same claim—the prediction made on Tuesday that it will be raining on Wednesday, say—lacks a truth-value when assessed relative to the time of utterance (Tuesday), but is either true or false when assessed relative to the time whose goings on the claim is making a prediction about (Wednesday).

But as Elizabeth Barnes and I have argued,[28] if you think openness consists in indeterminacy, and you think indeterminacy is compatible with bivalence, then there is no problem in accounting for this data concerning retrospective assessment even if, *contra* MacFarlane, you are an absolutist about the truth-value of tensed claims. To say that my earlier prediction was open is, on this view, simply to say that it was neither determinately true nor determinately false. But indeterminacy, on the model defended by Barnes and by Barnes and Williams and that I am adopting here, does not exclude truth or falsity. If p is indeterminate, it is either true or false, it is simply indeterminate which. Thus I can look out at the rain now and say that my prediction yesterday that it would rain today was indeed true when I made it, and this is not inconsistent with my nevertheless claiming that the future was open with respect to that prediction. Openness, being indeterminacy, does not exclude truth or falsity. "My prediction was true, although it was open what would happen" amounts to "My prediction was true, although it was not determinate that it was true," which is perfectly consistent given a bivalent account of indeterminacy.

Reason 3: The Indeterminacy Account Coheres Better with our Cognitive Attitudes towards Future Contingents

Suppose you are completely confident that the future is open with respect to whether it will rain tomorrow. Suppose further that you are completely confident that openness consists in there being no fact of the matter. Putting these together, you become supremely confident that there is no fact of the matter whether it will rain tomorrow; that is, you are supremely confident that "It will rain tomorrow" is neither true nor false. *Prima facie*, this means you ought to completely reject both the claim that it will rain tomorrow and its negation. If I am completely confident that some claim is not true, I ought to completely reject that claim. But this seems wrong: my confidence in it being open whether it will rain should not lead me to completely reject the claim that it will rain.

[27] MacFarlane (2003, 2008). [28] Barnes and Cameron (2009, pp297–8).

Quite the reverse: my confidence that it is *open* should lead me *not* to rule that option out.[29]

The point here is that our cognitive attitudes towards future contingents do not seem to be like our cognitive attitudes towards paradigm cases of there being no fact of the matter. If you are convinced that strict finitism is true and that mathematical reality is thereby not extensive enough to speak to either the truth or the falsity of the continuum hypothesis, then it seems absolutely right that you ought to completely reject both the continuum hypothesis and its negation. Both that claim and its negation describe reality as being a way that you are completely convinced it is not, so you ought to utterly reject each. By contrast, even if I am completely convinced that it is open what tomorrow's weather will be like, I am not thereby completely convinced that "It will rain tomorrow" and its negation misdescribe reality. Quite the opposite: I am convinced that one of them gets it right. I just think it is unsettled which.

This difference in appropriate cognitive attitudes between the cases manifests itself in a difference in how it is rational to act in each case. Suppose God tells me—and so I am utterly confident that this is true—that He is going to zap my house with lightning bolts if and only if the continuum hypothesis is true. If I have any credence at all that the continuum hypothesis is true, I ought to get out of the house, since there is not much disutility in evacuating, whereas being in the house while it is struck by lightning is very bad. But if I am really convinced that strict finitism is true, I need not do anything. I am totally confident that the continuum hypothesis is not true, so totally confident that God will not be zapping my house with lightning. Insofar as I worry about the lightning striking, that simply reflects my lack of confidence in the strict finitism whose truth rules out the truth of the continuum hypothesis. By contrast, suppose that a fair coin is due to be tossed at noon tomorrow, and I am utterly confident that the future is open with respect to whether or not it will land heads when tossed, and God tells me that He is going to zap my house with lightning if and only if it is true that it will land heads tomorrow. Intuitively, I think, I ought to evacuate. My belief that it is open what will happen does not conflict with, indeed it *mandates*, my thinking that the coin's landing heads is something that might happen. My belief that it is open what will happen does not force me to outright reject, indeed it renders it absurd to outright reject, "The coin will land heads." Planning on the possibility of it being the case that the coin will land heads in no way reflects a lack of confidence in it being open whether the coin will land heads in the way

[29] See Barnes and Cameron (2011), and see Williams (2014) for detailed discussion of related issues.

that planning on the possibility of the continuum hypothesis being true does reflect a lack of confidence in strict finitism. That, I suggest, is because it being open whether the coin will land heads does not exclude the truth of "The coin will land heads" in the way that the truth of strict finitism excludes the truth of the continuum hypothesis. This suggests that openness consists not in there being no fact of the matter what will happen, but in it being indeterminate what will happen. It is not that both options (that the coin will land heads and that it will land tails) are excluded—one of them correctly describes what is going to be the case—it is simply that it is unsettled *which* of them correctly describes what is going to be the case.

For these reasons, I prefer the account of the open future that says that (at least some) future contingent claims are indeterminate in truth-value over one that says that there is no fact of the matter concerning (at least some) future contingents. The view I accept, then, says that there is some way things will turn out, it is just indeterminate *how* things will turn out. There are multiple candidate complete future histories—descriptions of how the future will unfold in every detail—such that it is determinately the case that exactly one of those complete future histories is true—that it accurately describes how the future will unfold—but that it is indeterminate *which* of those multiple candidate complete future histories is true.

For it to be indeterminate which of those complete future histories is true, there must be a corresponding indeterminacy in the ontology that makes it the case that things will be a certain way. So to bring this all back to the moving spotlight metaphysic defended in chapter four, the thesis that the future is open demands that there be a certain amount of indeterminacy either in what future entities exist, or in the ages, distributional properties or locations that things have now, since once you have settled these facts you have settled all the truths about what will happen.

So if it is open, for example, whether little Anita will grow to 5ft tall or 5.5ft tall, this is a matter of it being indeterminate what temporal distributional property Anita has now (i.e. which one she has simpliciter). There will, of course, be plenty of temporal distributional properties that Anita determinately does not have, such as ones that entail that their bearer is 6ft tall at some point during its life, and ones that entail that their bearer is at no point during its life taller than 4ft. But there will be more than one temporal distributional property such that it is determinate that exactly one of those is the temporal distributional property that Anita has now, but it is indeterminate *which* of those properties is the one that Anita has now. One of those candidate distributional properties has Anita's height vary over her life to a maximum of 5ft, the other has her height vary over her life to a maximum of 5.5ft. Since for each of these properties it is indeterminate whether Anita instantiates that property, it is thereby indeterminate how her height will vary over her life. It is determinately true that she will grow to over 4ft tall, and determinately false that she will grow to 6ft tall, but it is indeterminate whether she will grow to 5ft tall or to 5.5ft tall. And so the future is open with respect to what maximum height Anita will grow to. The future contingent "Anita will be 5.5ft tall" is open: it is determinately either true or false, but it is indeterminate which.

Or suppose it is open whether a coin will land heads or tails. This will amount to an indeterminacy in the location relation the coin now stands in. For the coin to land heads requires that it be located one way in spacetime, for it to land tails requires that it be located another way. Determinately, the coin is located one way or the other, but it is indeterminate which, and as a result it is determinate that the coin will either land heads or tails, but it is indeterminate which, and so the future is open with respect to how the coin will land. "The coin will land heads" is neither determinately true nor determinately false, although it is determinately one or the other.

Or suppose it is open whether or not Ahmed and Julianne will have a child. This will amount to an indeterminacy in what future entities there are: that is, in an indeterminacy in what there is, simpliciter. There will be candidate domains for the unrestricted existential quantifier that include a thing that is a future child of Ahmed and Julianne and candidate domains that do not include any such thing. It is determinately the case that exactly one of these candidate domains is the correct one—the one that includes all and only the things that there are—but it is indeterminate which of these candidate domains is the unique correct one. Hence it is indeterminate whether one of the things that there unrestrictedly are is a future child of Ahmed and Julianne; hence it is indeterminate whether Ahmed and Julianne will have a child; hence the future is open as to whether Ahmed and Julianne will have a child. "Ahmed and Julianne will have a child" is, determinately, either true or false, but it is not determinately true and it is not determinately false.

Return now to the growing blocker's thought from section 5.1. The argument for the growing block is that the metaphysical asymmetry between the fixed past and open future requires an ontological asymmetry: that the past be real but the future unreal. I think that has been revealed as a mistake. The metaphysical asymmetry can be accounted for simply by the extent to which there is indeterminacy in the world. The past is fixed but the future open simply because there are various ways for the world to be such that the world is determinately one of those ways, and how things will be depends on which of those ways is the way the world is, whereas how things were is the same no matter which of those ways is the way the world is.

For example, it is indeterminate what temporal distributional property Anita has. Some temporal distributional properties that are not determinately lacked by her describe her height as varying in such a way that it reaches 5.5ft, others describe her height as varying in such a way that she is never taller than 5ft. But each of the temporal distributional properties such that it is not determinate that Anita lacks them describes Anita's height as starting from 1.6ft (say). Hence, while it is open what Anita's height will be in the future, it is fixed what it has been in the past; it is indeterminate whether Anita will ever be 5.5ft tall, but it is determinate that she was 1.6ft tall, because while it is indeterminate what temporal distributional property she has, it is determinate that the temporal distributional property that she has describes her height as starting at 1.6ft tall. Likewise, *mutatis mutandis*, for all historical truths. While it might be indeterminate *what* makes them true, it is determinate that *something* makes them true, because the various candidate options for how the world might be all agree on how the past was, they just disagree on how the future is.

The growing blocker might ask what right the moving spotlighter has to assume that the extent of indeterminacy in the world is just enough to ensure that the future is open whilst securing the fixity of the past. Answer: this is simply a hypothesis about how the world is—a hypothesis that is *justified* because it secures our pre-theoretic intuition that the past is fixed and the future open. I think it is perfectly reasonable to accept an ontological hypothesis on such grounds. And if it is not—well, so much the worse for the growing blocker as well, for their characteristic claim about ontology, that the past is real and the future unreal, is also simply an ontological hypothesis whose sole justification is that it secures the fixity of the past and the openness of the future. Either it is acceptable to use the intuition about this metaphysical difference between the past and the future to guide you in your metaphysics, or it is not: the growing blocker does not have any advantage over the moving spotlighter here. And if what I said above is right, that the moving spotlighter's account of openness in terms of indeterminacy is better than the growing blocker's account of openness as there being no fact of the matter, then the moving spotlighter is at an advantage when it comes to the open future.

One final issue. If we account for openness by postulating an indeterminacy in ontology, this motivates a refinement to truthmaker theory as that view is stated in section 3.2. Truthmaker theory, as I characterized it, is the view that the fundamental truths are all truths solely about what there is. I glossed this in that section as saying that the fundamental account of how reality is will simply be of the form "A exists, B exists, C exists... and that's all that exists." I think believing in the open future in the sense defended in this chapter should lead us to say that the fundamental account of reality instead tells us the various candidate *options* concerning what there is. That is, the fundamental truths will be of the form "Determinately, A exists," "It's indeterminate whether B exists," "Determinately A exists or B exists, but not both," etc. This *is* a departure from truthmaker theory as stated in chapter three, but I think it is an entirely motivated one, and still consistent with the motivations for truthmaker theory. The guiding thought behind truthmaker theory, as developed in chapter three, is that the fundamental truths are all truths about what there is: truths that merely characterize what belongs in the ontological inventory. These truths *do* do that: to say that A determinately exists, or that it is indeterminate whether B exists, *is* to characterize the ontological inventory. These claims *are* simply about what there is, and so I think allowing such truths as fundamental is consistent with the motivation behind truthmaker theory.

In this respect I think determinacy is different from, say, tense or modality. It is consistent with the goals of truthmaker theory to take as fundamental truths about what there determinately is, or truths about it being indeterminate whether something exists, whereas I do not think it would be consistent with the goals of truthmaker theory to take as fundamental truths about what there was or always has been, or about what there could be or what there necessarily is. That is why I think the truthmaker theorist is obliged to accept something like the truth-making account for historical truths that is put forward in chapter four, rather than taking truths about what existed or will exist as fundamental even if, in accepting the account of the open future defended in this chapter, she takes as fundamental truths about what there determinately is, or about what ontologies the world is indeterminate between.

Why this difference? Why is it okay for the truthmaker theorist to accept "Determinately, A exists" but not "It was the case that A exists" as a fundamental truth about the world? I think that claims about what there (in)determinately is characterize the ontological inventory in a way that claims about what there was or could be do not—they only characterize how the ontological inventory used to be, or how it could have been. If you ask me whether there are Xs and I reply "It's indeterminate," that is a perfectly good answer, and you should be satisfied that that is the best answer I could have given. But if instead I reply "There were" or "There could have been" you should not be satisfied—you should respond: "I didn't ask whether there *were/could have been* any Xs, I asked whether there *are* any!" And that is because my answer that it is indeterminate whether there are Xs speaks to how the ontological inventory in fact is. It *is* a claim about what there is.

Whereas saying that there were/could be Xs leaves open the question of how the ontological inventory in fact *is* (actually, now). Those are *not* claims about what there is, they are merely claims about what there used to be or could have been. That is why the latter cannot be fundamental, according to truthmaker theory: truths about what there used to be or could have been must, if truthmaker theory is true, be true in virtue of truths about what there in fact is. The ontological inventory being as it is must settle how the ontological inventory was or could have been (since it must settle *everything*). Whereas truths about what there determinately is, etc., can be fundamental, because they characterize how the ontological inventory is. Tense and modality are used to describe *alternatives* to reality: we use tensed and modal language to state claims not about how things are but about how they used to be or could be. Whereas to say that something is determinately true, or that it is indeterminate what is the case, is to say something about how things in fact are, not to describe some alternative to how things are. That is why modifying an existential claim with a determinacy operator is still to characterize the ontological inventory as it in fact is, whereas to modify an existential claim with a tense or modal operator is merely to characterize an alternative to our ontology—to say what used to be our ontology or what could have been our ontology. And so I think accepting the account of the open future proposed in this chapter, while it mandates a *revision* of truthmaker theory, is still within the spirit of the theory; and there is still good reason for the truthmaker theorist to demand an ontological underpinning of tensed truths of the kind that was offered in the previous chapter.

Conclusion

Let me take a step back to offer some concluding thoughts. The most familiar metaphysics of time are B-Theoretic eternalism and presentism. Both the B-Theorist and the presentist agree that there is such a thing as the way things are simpliciter, but they disagree on what kind of description of the world is needed to describe the way things are simpliciter. For the B-Theorist the way things are simpliciter is an *atemporal* description of how the world is *across* time, whereas for the presentist the way things are simpliciter is an *instantaneous* description of how the world is *now*.

It is tempting to think of the eternalist A-Theorist (whether the moving spotlighter or growing blocker) as being committed to giving both descriptions. In believing in the reality of the non-present, she believes that the world extends across time, and so has to give an atemporal description of how things are across time in order to say how things are simpliciter. But in believing in a privileged present, she believes in a time such that what is happening now amounts to what is happening at that time, so surely the description of how things are simpliciter ought to also include the instantaneous description of how things stand at that time.

But these two descriptions of the world contradict one another! In describing the world from the atemporal perspective we will say that there are dinosaurs. In describing it from the perspective of the present, we will say that there are no dinosaurs. So what is true? Are there dinosaurs or not?

One solution is to say that this question is not well formed, because there simply is no such thing as truth simpliciter. We can ask whether it is *atemporally* true that there are dinosaurs (answer: yes), and we can ask whether it is *now* true that there are dinosaurs (answer: no), and that is it: there is no asking whether it is true *simpliciter*. I think this should be the option of last resort. A guiding assumption for me that I will not give up unless I am forced to is that there is such a thing as truth, simpliciter: that we can ask "How are things?," "What is the case?," and expect an answer, without having to ask "How are things *now*?" or "What is the case, atemporally speaking?"

Another option is to grant that there is such a thing as truth simpliciter and say that both the atemporal description of how things are across time and the temporal description of how things are presently are correct descriptions of how the world is, but argue that they cannot be given together.[1] This would be to hold that you can say that there are dinosaurs and be giving a complete account of reality that is true simpliciter, and you can say that there are no dinosaurs and be giving a complete account of reality that is true simpliciter, but you cannot ever say that there are dinosaurs and that there are no dinosaurs. Each perspective gives a true and complete account of the world, but you can only describe the world from one perspective at a time. This is a very odd view. There is truth simpliciter, but there is no *unique* description of the world that is true simpliciter. There are rival complete descriptions of reality that contradict one another but that are each true from a certain perspective on reality, neither of which is better than the other.[2]

If one wishes to be an eternalist A-Theorist whilst holding on to the massively intuitive idea that there is one unique complete description of reality that is true simpliciter, then one has to privilege one of the descriptions: either the atemporal one or the one from the present perspective. The way we have become used to thinking about views like the moving spotlight—indeed, the way that is suggested by the metaphor—is to privilege the atemporal perspective. Reality is the way it is spread out across time; the past, present, and future are equally real, and things are spread out through, and vary across, time. Reality is just the way the B-Theorist thinks it is, in fact, but for an extra feature: one time is special. That atemporal description of things—that things are as the B-Theorist says they are but with an extra addition describing the presentness of one time—is how things are, simpliciter. And of course, how things are simpliciter *changes*, but only with respect to *which* time it is that is special.[3]

Saying that something is *presently* true, on this view, is to speak restrictedly: it is to speak with one's attention restricted to the goings on at the time under the spotlight. So it is presently true that there are no dinosaurs, because no dinosaurs are under the spotlight. This moving spotlighter agrees with the B-Theorist that to say how things are now is to give an *incomplete* description of reality—it is merely to describe a *portion* of reality. The only difference is that while the B-Theorist thinks that the relevant portion we are restricting our attention to is picked out indexically—it is the portion at which we are having *this* very

[1] See Sider (2013a, pp272–3). [2] Cf. Fine's Fragmentalism, discussed in §2.4.
[3] This is how Sider characterizes the moving spotlight view in Sider (2013a, pp259–1).

thought—the moving spotlighter thinks that the portion is picked out by some objective feature—it is the portion that has the special property of presentness.

The main thesis of this book is that this is a bad way of understanding the moving spotlight view, and that the moving spotlighter ought instead to privilege the description of reality from the present perspective.

One way in which the view that privileges the atemporal description is bad is that it does not seem to really do justice to the thought that there is something special about the present. Just as for the B-Theorist, facts about what is happening now are mere incomplete descriptions of reality on this metaphysic. Sure, there are no dinosaurs now. But there *are* dinosaurs, simpliciter. It is the latter that tells you about how reality stands with respect to that existential question; the former only gives you a partial story concerning how things stand dinosaur-wise. The only difference between the B-Theory and this version of the moving spotlight theory is that for the latter the restriction on reality that gives you the incomplete description is given by some objective feature rather than something picked out indexically. But why should that make the description any better a description of reality? I could restrict my attention to what is happening within a meter of an electron, and so when I ask "Are there Fs?" I am asking whether any Fs are located within a meter of an electron. The restriction on reality is given by an objective feature—spatial proximity to something instantiating electronhood—but that does not make the restriction any more relevant to a metaphysician interested in whether Fs exist. Why, then, should the restriction to what is present be relevant, when that is not the entirety of how things are? For this moving spotlighter and for the B-Theorist alike, the description is but an incomplete account of reality. Why should an incomplete account of reality be relevant to us as metaphysicians? Why are the present goings on more *important* than past and future goings on on this account? Another way in which this view is bad is that the only real change is the moving of the spotlight: that is, the changing of which portion of reality is the one that we restrict our attention to when saying what is true now. How ordinary things are is as static on this view as it is on the B-Theory. (See chapter three.)

The moving spotlighter, I suggest, ought instead to follow the presentist's lead and privilege the description of reality from the present perspective. How things are *simpliciter* is how things are *now*. In giving a description of how things are from the present perspective, we are giving a complete account of how reality is. That is why present goings on are more important than past and future goings on: just as for the presentist, the present goings on exhaust reality for this moving spotlighter. And there is genuine change in all the ordinary goings on, on this metaphysic, just as much as there is for the presentist.

The difference between this moving spotlighter and the presentist is just this: that the moving spotlighter grants that one can speak from the present perspective *about* the non-present. That one can say how non-present things *now* are. Truth simpliciter is present truth, but amongst the way things are now—*contra* presentism—is that mere past and future entities are some way or other.

This moving spotlighter, then, does *not* believe in the reality of the past and future. The way things were and will be is no part of the complete description of how reality is. What she believes in is the reality of past and future *things*. She believes in Caesar and the first lunar colony as well as the Scottish parliament and Bruce Springsteen. But for each of these four things, they can only be truly described by saying how they are *now*. The way they were and will be is no part of reality.

So how are Caesar and the first lunar colony *now*? On that, details will vary. I have defended a particular view on the properties of things in chapter four, but even if my particular development of this kind of moving spotlight metaphysic is rejected, I hope that I can convince people to pay more attention to this way of thinking about the moving spotlight view. Think of the moving spotlight not as an enriched B-Theory—as taking the atemporal description of the world and making one moment of it special—but rather as an enriched presentism—as taking the present-tensed description of the world and allowing it to describe non-present things.

Bibliography

Alchourrón, Carlos and Bulygin, Eugenio, (1971), *Normative Systems*, New York: Springer Verlag.

Armstrong, David, (1997), *A World of States of Affairs*, Cambridge: Cambridge University Press.

Armstrong, David, (2006), 'Reply to Heil', *The Australasian Journal of Philosophy* 84(2), 245–7.

Barnes, Elizabeth, (2006), *Conceptual Room for Ontic Vagueness*, PhD Thesis, University of St Andrews.

Barnes, Elizabeth, (2010), 'Ontic Vagueness: A Guide for the Perplexed', *Noûs* 44(4), 607–27.

Barnes, Elizabeth, (2013), 'Metaphysically Indeterminate Existence', *Philosophical Studies* 166(3), 495–510.

Barnes, Elizabeth and Cameron, Ross, (2009), 'The Open Future: Bivalence, Determinism and Ontology', *Philosophical Studies* 146(2), 291–309.

Barnes, Elizabeth and Cameron, Ross, (2011), 'Back to the Open Future', *Philosophical Perspectives* 25, 1–26.

Barnes, Elizabeth and Williams, J. R. G., (2011), 'A Theory of Metaphysical Indeterminacy', *Oxford Studies in Metaphysics* 6, 103–48.

Baxter, Donald and Cotnoir, Aaron, (2014), *Composition as Identity*, Oxford: Oxford University Press.

Bigelow, John, (1996), 'Presentism and Properties', *Noûs* 30, Supplement: Philosophical Perspectives 10, Metaphysics.

Blackburn, Simon, (1993), *Essays in Quasi-Realism*, Oxford: Oxford University Press.

Bourne, Craig, (2002), 'When am I? A Tense Time for Some Tense Theorists?', *Australasian Journal of Philosophy* 80(3), 359–71.

Bourne, Craig, (2006), *A Future For Presentism*, Oxford: Oxford University Press.

Braddon-Mitchell, David, (2004), 'How Do We Know It Is Now Now?', *Analysis* 64(3), 199–203.

Bricker, Phillip, (2006), 'Absolute Actuality and the Plurality of Worlds', *Philosophical Perspectives* 20(1), 41–76.

Broad, C. D. (1923), *Scientific Thought*, London: Routledge and Keegan Paul Ltd.

Cameron, Ross, (2007), 'The Contingency of Composition', *Philosophical Studies* 136(1), 99–121.

Cameron, Ross, (2008a), 'Turtles all the Way Down: Regress, Priority and Fundamentality', *The Philosophical Quarterly* 58(230), 1–14.

Cameron, Ross, (2008b), 'How to Be a Truthmaker Maximalist', *Noûs* 42(3), 410–21.

Cameron, Ross, (2008c), 'Truthmakers and Modality', *Synthese* 164(2), 261–80.

Cameron, Ross, (2011), 'Truthmaking for Presentists', *Oxford Studies in Metaphysics* 6, 55–100.

Cameron, Ross, (2012), 'Why Lewis's Analysis of Modality Succeeds in its Reductive Ambitions', *Philosophers' Imprint* 12(8), 1–21.

Cameron, Ross, (forthcoming), 'Truthmakers', in Michael Glanzberg (ed.), *The Oxford Handbok of Truth*, Oxford: Oxford University Press.

Cappelen, Herman and Hawthorne, John, (2010), *Relativism and Monadic Truth*, Oxford: Oxford University Press.

Caplan, Ben and Sanson, David, (2010), 'The Way Things Were', *Philosophical and Phenomenological Research* 81(1), 24–39.

Chihara, Charles, (1998), *The Worlds of Possibility*, Oxford: Oxford University Press.

Craig, William Lane, (1998), 'McTaggart's Paradox and the Problem of Temporary Intrinsics', *Analysis* 58(2), 122–7.

Diekemper, Joseph, (2005), 'Presentism and Ontological Symmetry', *The Australasian Journal of Philosophy* 83(2), 223–40.

Egan, Andy, (2010), 'Disputing About Taste', in Richard Feldman (ed.), *Disagreement*, Oxford: Oxford University Press, 247–92.

Ehring, Douglas, (1997), 'Lewis, Temporary Intrinsics, and Momentary Tropes', *Analysis* 57, 254–8.

Evans, Gareth, (2002), 'Does Tense Logic Rest on a Mistake?', in his *Collected Papers*, Oxford: Oxford University Press, 343–63.

Fine, Kit, (1994), 'Essence and Modality', *Philosophical Perspectives* 8, 1–16.

Fine, Kit, (2000), 'The Question of Realism', *Philosophers' Imprint* 1(2), 1–30.

Fine, Kit, (2005), 'Tense and Reality', in his *Modality and Tense: Philosophical Perspectives*, Oxford: Oxford University Press, 261–320.

Fine, Kit, (2012), 'Guide to Ground', in Fabrice Correia and Benjamin Schneider (eds.), *Metaphysical Grounding*, Cambridge: Cambridge University Press, 37–90.

Fine, Kit, (2013), 'Fundamental Truth and Fundamental Terms', *Philosophy and Phenomenological Research* 87(3), 725–32.

Forrest, Peter, (2004), 'The Real But Dead Past: a Reply to Braddon-Mitchell', *Analysis* 64, 358–62.

Greenough, Patrick, (2008), 'Indeterminate Truth', in Peter French (ed.), *Truth and Its Deformities, Midwest Studies in Philosophy* 32(1), 213–41.

Hale, Bob and Wright, Crispin, (2001), *The Reason's Proper Study*, Oxford: Oxford University Press.

Haslanger, Sally, (1989), 'Endurance and Temporary Intrinsics', *Analysis* 49, 119–25.

Hawley, Katherine, (2001), *How Things Persist*, Oxford: Oxford University Press.

Hazen, Allen, (1979), 'Counterpart-Theoretic Semantics for Modal Logic', *The Journal of Philosophy* 76(6), 319–88.

Heathwood, Christopher, (2005), 'The Real Price of the Dead Past', *Analysis* 65(3), 249–51.

Hiller, Avram and Neta, Ram, (2007), 'Safety and Epistemic Luck', *Synthese* 158, 303–13.

Hudson, Hud and Wasserman, Ryan, (2010), 'van Inwagen on Time Travel and Changing the Past', *Oxford Studies in Metaphysics* 5, 41–9.

Jacobs, Jonathan, (2010), 'A Powers Theory of Modality: Or, How I Learned To Stop Worrying and Reject Possible Worlds', *Philosophical Studies* 151(2), 227–48.

Jubien, Michael, (1988), 'Problems With Possible Worlds', *Philosophical Analysis* 39, 299–322.

King, Jeffrey C., (2003), 'Tense, Modality, and Semantic Values', *Philosophical Perspectives* 17, 195–245.

King, Jeffrey, (2007), *The Nature and Structure of Content*, New York: Oxford University Press.

Kripke, Saul, (1980), *Naming and Necessity*, Cambridge, MA: Harvard University Press.

Kripke, Saul, (1982), *Wittgenstein on Rules and Private Language*, Oxford: Blackwell.

Lewis, David, (1973), *Counterfactuals*, Oxford: Blackwell.

Lewis, David, (1983), 'New Work for a Theory of Universals', *The Australasian Journal of Philosophy* 61, 343–77.

Lewis, David, (1986), *On the Plurality of Worlds*, Oxford: Blackwell.

MacFarlane, John, (2003), 'Future Contingents and Relative Truth', *The Philosophical Quarterly* 53, 321–36.

MacFarlane, John, (2008), 'Truth in the Garden of Forking Paths', in Max Kölbel and Manuel García-Carpintero (eds), *Relative Truth*, Oxford: Oxford University Press, 81–102.

McCall, Storrs, (1994), *A Model of the Universe: Space-Time, Probability, and Decision*, Oxford: Clarendon Press.

McDaniel, Kris, (2009), 'Extended Simples and Qualitative Hetereogeneity', *The Philosophical Quarterly* 59(235), 325–31.

McTaggart, J. Ellis, (1908), 'The Unreality of Time', *Mind* 17(68), 457–74.

Manley, David, (2007), 'Safety, Content, Apriority, Self-Knowledge', *The Journal of Philosophy* 104, 403–23.

Mares, Edwin, (2004), 'Semantic Dialetheism', in G. Priest, J. C. Beall, and B. Armour-Garb (eds), *The Law of Non-Contradiction: New Philosophical Essays*, Oxford: Oxford University Press, 264–75.

Markosian, Ned, (1995), 'The Open Past', *Philosophical Studies* 79, 95–105.

Markosian, Ned, (2004), 'A Defense of Presentism', *Oxford Studies in Metaphysics* 1, 47–82.

Markosian, Ned, (2014a), 'Time', *The Stanford Encyclopedia of Philosophy* (Spring 2014 Edition), Edward N. Zalta (ed.), URL = <http://plato.stanford.edu/archives/spr2014/entries/time/>.

Markosian, Ned, (2014b), 'The Truth About the Past and the Future', in Fabrice Correia and Andrea Iacona (eds), *Around the Tree: Semantic and Metaphysical Issues Concerning Branching Time and the Open Future*, Springer, 127–41.

Martin, C. B., (1996), 'How It Is: Entities, Absences and Voids', *The Australasian Journal of Philosophy* 74, 57–65.

Maudlin, Tim, (2007), *The Metaphysics Within Physics*, Oxford: Oxford University Press.

Meinong, Alexius, (1915), *On Possibility and Probability. Contributions to Object Theory and Epistemology*, Leipzig: JA Barth.

Mellor, D. H., (1998), *Real Time II*, London: Routledge.

Mellor, D. H., (2003), 'Real Metaphysics: Replies', in H. Lillehamer and G. Rodriguez-Pereyra (eds), *Real Metaphysics*, London: Routledge, 212–38.

Merricks, Trenton, (1999), 'Persistence, Parts and Presentism', *Noûs* 33, 421–38.

Merricks, Trenton, (2006), 'Good-Bye Growing Block', *Oxford Studies in Metaphysics* 2, 103–10.

Merricks, Trenton, (2007), *Truth and Ontology*, Oxford: Clarendon Press.

Merricks, Trenton, (2009), *Truth and Ontology*, Oxford: Oxford University Press.
Meyer, Ulrich, (2013), 'The Triviality of Presentism', in R. Ciuni, K. Miller, and G. Torrengo (eds), *New Papers on the Present*, Munich: Philosophia Verlag, 67–88.
Miller, Alexander, (1998), *Philosophy of Language*, London: Routledge.
Mumford, Stephen, (2007a), 'Negative Truth and Falsehood', *Proceedings of the Aristotelian Society* 107, 45–71.
Mumford, Stephen, (2007b), 'A New Solution to the Problem of Negative Truth', in J.-M. Monnoyer (ed.), *Metaphysics and Truthmakers*, Frankfurt: Ontos Verlag, 313–29.
Nolan, Daniel, (2008), 'Truthmakers and Predication', *Oxford Studies in Metaphysics* 4, 171–92.
Parsons, Josh, (2000), 'Must a Four-Dimensionalist Believe in Temporal Parts?', *The Monist* 83(3), 399–418.
Parsons, Josh, (2004), 'Distributional Properties', in Frank Jackson and Graham Priest (eds), *Lewisian Themes*, Oxford: Oxford University Press, 173–80.
Pasnau, Robert, (2011), *Metaphysical Themes 1274–1671*, Oxford: Clarendon Press.
Pasnau, Robert, (2014), 'Epistemology Idealized', *Mind* 122, 987–1021.
Paul, L. A., (2010), 'Temporal Experience', *The Journal of Philosophy* CVII(7), 333–59.
Perry, John, (1979), 'The Problem of the Essential Indexical', *Noûs* 13, 3–21.
Priest, Graham, (2006), *In Contradiction*, 2nd edition, Oxford: Oxford University Press.
Quine, W. V., (1951), 'Ontology and Ideology', *Philosophical Studies* 2(1), 11–15.
Quine, W. V., (1980), 'On What There Is', in his *From a Logical Point of View*, 2nd edition, Cambridge, MA: Harvard University Press, 1–19.
Schaffer, Jonathan, (2010), 'Monism: The Priority of the Whole', *The Philosophical Review* 119(1), 31–76.
Sider, Theodore, (2003a), *Four-Dimensionalism: An Ontology of Persistence and Time*, Oxford: Clarendon Press.
Sider, Theodore, (2003b), 'Against Vague Existence', *Philosophical Studies* 114, 135–46.
Sider, Theodore, (2007a), 'NeoFregeanism and Quantifier Variance', *Aristotelian Society Supplementary Volume* 81, 201–32.
Sider, Theodore, (2007b), 'Parthood', *The Philosophical Review* 116, 51–91.
Sider, Theodore, (2013a), *Writing the Book of the World*, Oxford: Oxford University Press.
Sider, Theodore, (2013b), 'Replies to Dorr, Fine, and Hirsch', *Philosophy and Phenomenological Research* 87(3), 733–54.
Sider, Theodore, (ms.), 'Scientific Properties, and How Tools in Metaphysics Matter'.
Skow, Bradford, (2009), 'Relativity and the Moving Spotlight', *The Journal of Philosophy* 106, 666–78.
Smith, Nicholas J. J., (2011), 'Inconsistency in the A-Theory', *Philosophical Studies* 156, 231–47.
Soeteman, Arend, (1997), 'On Legal Gaps', in E. G. Valdez, W. Krawietz, G. H. von Wright, and R. Zimmerling (eds), *Normative Systems in Legal and Moral Theory*, Berlin: Duncker, 323–32.
Sullivan, Meghan, (2012), 'The Minimal A-Theory', *Philosophical Studies* 158(2), 149–74.
Sullivan, Meghan, (2014), 'Change We Can Believe In (and Assert)', *Noûs* 48(3), 474–95.
Tallant, Jonathan, (2012), 'Time for Distribution?', *Analysis* 72, 264–70.

Tallant, Jonathan, and Ingram, David, (2012), 'Presentism and Distributional Properties', *Oxford Studies in Metaphysics* 7, 305–14.

Tooley, Michael, (1997), *Time, Tense, and Causation*, Oxford: Clarendon Press.

Van Cleve, James, (1996), 'If Meinong is Wrong, is McTaggart Right?', *Philosophical Topics* 24(1), 231–54.

van Inwagen, Peter, (1985), 'Plantinga on Trans-World Identity', in J. E. Tomberlin and P. van Inwagen (eds), *Alvin Plantinga*, Dordrecht: D. Reidel Publishing Company, 101–20.

van Inwagen, Peter, (1990), 'Four-Dimensional Objects', *Noûs* 24, 245–55.

van Inwagen, Peter, (1995), *Material Beings*, Ithaca, NY: Cornell University Press.

van Inwagen, Peter, (2010), 'Changing the Past', *Oxford Studies in Metaphysics* 5, 3–28.

Vetter, Barbara, (2015), *Potentiality: From Dispositions to Modality*, Oxford: Oxford University Press.

Wallace, Megan, (2011), 'Composition as Identity: Part 1', *Philosophy Compass* 6(11), 804–16.

Wang, Jennifer, (forthcoming), 'The Modal Limits of Dispositionalism', *Noûs*.

Williams, J. Robert G., (2014), 'Decision Making Under Indeterminacy', *Philosophers' Imprint* 14(4), 1–34.

Williamson, Timothy, (2000), *Knowledge and its Limits*, Oxford: Clarendon Press.

Williamson, Timothy, (2002), 'Necessary Existents', in A. O'Hear (ed.), *Logic, Thought and Language*, Cambridge: Cambridge University Press, 233–51.

Williamson, Timothy, (2013), *Modal Logic as Metaphysics*, Oxford: Oxford University Press.

Wilson, Jessica, (2013), 'A Determinable-Based Account of Metaphysical Indeterminacy', *Inquiry* 56(4), 359–85.

Woodward, Richard, (2011), 'Metaphysical Indeterminacy and Vague Existence', *Oxford Studies in Metaphysics* 6, 183–97.

Wright, Crispin, (1983), *Frege's Conception of Numbers as Objects*, Aberdeen: Aberdeen University Press.

Wright, Crispin, (1992), *Truth and Objectivity*, Cambridge, MA: Harvard University Press.

Wright, Crispin, (2001), *Rails to Infinity*, Cambridge, MA: Harvard University Press.

Wright, Crispin, (2003), 'Vagueness: A Fifth Column Approach', in J. C. Beall (ed.), *Liars and Heaps*, Oxford: Oxford University Press, 84–105.

Zimmerman, Dean, (1996), 'Persistence and Presentism', *Philosophical Papers* 25(2), 115–26.

Zimmerman, Dean, (2007), 'The Privileged Present: Defending an "A-Theory" of Time', in Ted Sider, John Hawthorne, and Dean W. Zimmerman (eds), *Contemporary Debates in Metaphysics*, Malden, MA: Blackwell, 211–25.

Index